*Advances in*
# MARINE BIOLOGY
## VOLUME 21

*Advances in*

# MARINE
# BIOLOGY

## VOLUME 21

*Edited by*

## J. H. S. BLAXTER

*Dunstaffnage Marine Research
Laboratory, Oban, Scotland*

## SIR FREDERICK S. RUSSELL

*Reading, England*

*and*

## SIR MAURICE YONGE

*Edinburgh, Scotland*

Academic Press                                        1984

(*Harcourt Brace Jovanovich, Publishers*)

London   Orlando   San Diego   San Francisco   New York
Toronto   Montreal   Sydney   Tokyo   São Paulo

ACADEMIC PRESS INC. (LONDON) LTD.
24–28 OVAL ROAD
LONDON NW1 7DX

*U.S. Edition published by*
ACADEMIC PRESS INC.
111 FIFTH AVENUE
NEW YORK, NEW YORK 10003

*British Library Cataloguing in Publication Data*

Advances in marine biology.—Vol. 21
1. Marine biology
574.92′05    QH91.A1

ISBN 0-12-026121-9

ISSN: 0065-2881

Typeset by Bath Typesetting Ltd., Bath
and printed by Thomson Litho Ltd., East Kilbride, Scotland

# CONTRIBUTORS TO VOLUME 21

F. E. RUSSELL, *College of Pharmacy, University of Arizona, Tucson, Arizona 85721, U.S.A.*

R. F. VENTILLA, *Sea Fish Industry Authority, Marine Farming Unit, Ardtoe, Acharacle, Argyll, PH36 4LD, Scotland.*\*

\*Currently working at ICLARM, MCC PO Box 1501, Makati, Metro-Manila, Philippines.

# SIR FREDERICK RUSSELL, 1897-1984

As Volume 21 was going to press we heard the sad news of Sir Frederick Russell's death on 5 June. He was, of course, the founder editor of *Advances in Marine Biology*. Until very shortly before his death, he maintained a very active interest in the production of the more recent volumes—reading proofs, commissioning reviews and corresponding with Academic Press and his co-editors. His knack of combining persuasion, firmness where necessary, and his great personal charm and friendliness greatly helped to ensure a regular flow of volumes. In his retirement he maintained his contacts with the scientific community and a full awareness of new developments. In 1976 at the age of 78, he produced a definitive work, *The Eggs and Planktonic Stages of British Marine Fishes*.

Freddie Russell had a long and distinguished career. He received his first degree at Gonville and Caius College, Cambridge. During the First World War he served in the Royal Naval Air Service and Royal Air Force, and was awarded the D.S.C., D.F.C. and Croix de Guerre. In 1922 he was appointed Assistant Director of Fisheries Research to the Egyptian Government. Shortly after, he joined the Marine Biological Association of the United Kingdom as an Assistant Naturalist and remained in Plymouth for 41 years. His first research, which attracted world renown, was on the diurnal vertical migration and distribution of ichthyoplankton and the crustacean copepods which are important as the food of fishes. In 1928, accompanied by his wife, Gweneth, he joined the Great Barrier Reef Expedition and worked on the physical, chemical and biological changes occurring there throughout the year. On his return to Plymouth he studied the biology of *Sagitta* and showed that different plankton "indicator" species were associated with particular water masses. These indicator species helped to explain the long-term changes in the ecosystem of the English Channel, now termed the Russell cycle.

During the Second World War he re-joined the RAF as an intelligence officer and, subsequently, returned to the MBA as Director. There he supervised post-war rebuilding, and by careful choice of staff and the expansion of facilities, he built up the laboratory to its present preeminent position in world marine biology. In spite of these onerous administrative duties, he was also able to write and illustrate a definitive monograph of the British Medusae, published in two large volumes by Cambridge University Press.

He also played a major part in the advance of marine biology outside the Plymouth laboratory—in the International Council for the Exploration of the Sea, in the Colonial Fisheries Advisory Committee, with the Central Electricity Generating Board and National Maritime Museum.

He received many honours: F.R.S. in 1938; C.B.E. in 1955; and a knighthood in 1965; honorary degrees from Birmingham, Bristol, Exeter and Glasgow Universities; and foreign membership of the Danish Academy. He was especially pleased when one of the Natural Environment Research Council's research vessels was named after him.

During his career he was greatly supported by his wife, who died in 1979. He never quite recovered from this loss but with characteristic decisiveness ran down his collection of books and journals, sold his house, and moved from Plymouth to Reading to be nearer his son.

His friends and colleagues will remember so many good qualities unusually combined into one person—courage, resilience in adversity, judgement, friendliness to young and junior scientists, and an old world courtesy. His flair for research and his role in marine biology over the last 50 years made him an outstanding and irreplaceable personality.

15 June 1984

J. H. S. Blaxter
A. J. Southward
C. M. Yonge

# CONTENTS

## Recent Developments in the Japanese Oyster Culture Industry

R. F. VENTILLA

# Marine Toxins and Venomous and Poisonous Marine Plants and Animals

F. E. RUSSELL

# Recent Developments in the Japanese Oyster Culture Industry

## R. F. Ventilla

*International Center for Living Resources Management MCC P.O. Box 1501,
Makati, Metro Manila, Philippines*

ADVANCES IN MARINE BIOLOGY VOL 21
ISBN 0–12–026121–9

# I. Introduction

Evidence of Japan's involvement with the oyster as a major source of food goes back several thousand years and can be seen in excavations of pre-historic middens all along the coast. The actual culture of oysters probably dates back to the beginning of the seventeenth century (early Tokugawa period) when it is recorded that oyster spat settling on stones and bamboo poles of fish traps were replanted in the intertidal zone surrounded by bamboo fences to keep predators out, which in turn acted as collectors of oyster spat. All records indicate that oyster culture began in Hiroshima Prefecture where Hiroshima Bay provided ideal sheltered conditions with a wide mud free shoreline exposed by an average 2·5 m tide, and near the ready market of the city of Japan's new thriving merchant class, Osaka. Until the 1900s three kinds of bottom culture developed in the different areas of Japan, viz: (a) rock culture, the oldest form of cultch, which was used on hard bottoms, (b) stick culture ("Kusatsu") which began in Hiroshima 300 years ago in which spat were collected on vertical bamboo poles arranged into fences and grown until spring of the following year when they were knocked off and scattered, (c) sowing culture on shores of sand and pebbles with seed from bamboo stick collectors. In this method the seed was scattered at low tide and levelled off by raking which broke shell margins and induced hardening of the shells.

The beginning of hanging culture is attributed to Seno and Hori of the Kanagawa laboratory (Tokyo University of Fisheries) who investigated hanging methods in 1923 (Seno and Hori, 1927; Seno, 1938). Following the success of these experiments, many new areas adopted this technique and this intensive method of hanging culture arose in other prefectures such as Shizuoka and Mie in south Honshu, Kumamoto on Kyushu island, and Iwate and Miyagi on the north-east coast of Honshu (see Fig. 1). Nowadays in Japan the culture of oysters is based on intensive off-the-bottom methods of three types (1) Rack culture (2) Raft culture and (3) Long-line culture (for details see pp. 19–29). The present industry in its steady production state, averaging over 200 000 tonnes in shell weight and 34 000 tonnes of meat per year from 1962, is an example of efficient utilization of a natural resource (see Table 1). The main obstacle to its further development is the insidious build up of pollutants (organic/inorganic) in vital areas such as Hiroshima Bay and the Seto inland sea in general, and the seeming reluctance of the cooperatives to accept maximum production thresholds with the dire consequences of ignoring such advice (see p. 26).

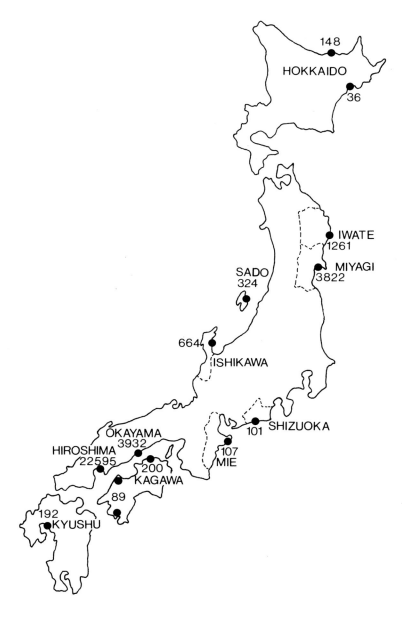

Fig. 1. The major oyster culture areas in Japan showing the quantity of oyster meat (tonnes) produced by the main prefectures in 1978.

TABLE I. JAPAN'S OYSTER PRODUCTION 1965–1978

| | National Production | | Hiroshima Production | |
| | In shell weight (tonnes) | Meat weight (tonnes) | Meat weight (tonnes) | % National |
|---|---|---|---|---|
| 1965 | 210 603 | 32 333 | 25 000 | 77·3 |
| 1966 | 221 102 | 35 313 | 27 500 | 77·8 |
| 1967 | 232 200 | 36 288 | 31 500 | 86·8 |
| 1968 | 265 881 | 41 530 | 31 188 | 75·1 |
| 1969 | 245 458 | 33 931 | 22 461 | 66·2 |
| 1970 | 169 752 | 26 515 | 19 358 | 54·2 |
| 1971 | 193 846 | 35 976 | 27 119 | 75·4 |
| 1972 | 217 373 | 33 107 | 23 479 | 70·1 |
| 1973 | 229 899 | 33 429 | 24 594 | 73·6 |
| 1974 | 210 583 | 33 111 | 23 712 | 71·6 |
| 1975 | 201 173 | 32 600 | 23 181 | 71·1 |
| 1976 | 226 286 | 34 292 | 22 500 | 65·6 |
| 1977 | 213 000 | 37 205 | 25 892 | 69·6 |
| 1978 | 232 000 | 33 935 | 22 595 | 66·6 |

# II. The Japanese Oyster

## A.  Characteristics

Seventeen species of oyster have been identified in Japanese waters although only five species of the genus *Crassostrea* have been utilized as food throughout history. Of these five species only *Crassostrea rivularis* (Gould) and particularly *Crassostrea gigas* (Thunberg) have been the subject of cultivation.

C. *rivularis* ("suminoe-gaki") is restricted to Western Kyushu, in the Ariake sea, although it is a good quality oyster and easily cultured. C. *gigas* ("magaki") is found everywhere in Japan from Hokkaido in the north, to Kyushu in the south, and occurs in several morphological variations, distinguishable by size, shape, colour, glycogen content and taste (Imai and Sakai, 1961). These regional differences are genetic, and cross breeding has produced strains of mixed or intermediate character. The main types are (a) the Hokkaido oyster which is large and thick shelled with a white mantle (specimens known up to 40 cm) (b) the Sendai or Miyagi oyster which is large and flat with a brown mantle and high water content and (c) the superior tasting Hiroshima oyster which is small, broad and deeply cupped with a black mantle and rapid growth characteristics. Crosses between the Hiroshima oyster and the Hokkaido variety produce a good tasting commercial oyster. Taste is one of these indefinable characteristics about which the Japanese are very serious. The Hiroshima oyster is the ultimate for the Japanese gastronome with the Miyagi oyster (and incidentally the European oyster *O. edulis*) considered second class. Taste in oysters is derived from their diet, and diatom feeding is considered essential for production of good tasting oysters. In the Hiroshima area they are now considering holding harvested oysters in tanks of monocultured diatoms to improve the flavour, since the diatom populations in the Bay have been depleted with the increasing chronic eutrophication. The Japanese oyster is remarkable for its nutritive value and an oyster based diet in the old days supplied the Japanese with more than 25 % of their daily protein requirements, as well as considerable glycogen, and 50 % of their calcium and phosphorus requirements together with all their iodine, iron and Vitamin A, B, C, D and G requirements (Cahn, 1950).

The oyster therefore represents one of the most nutritionally balanced items in the Japanese diet.

### B.   *Larval Development*

All the above species are found in subtropical to warm temperate conditions. Spawning and development therefore take place at water temperatures of around 20°C and higher, and the optimum temperature range for larval development is quoted as 23–25°C. *C. gigas* is oviparous and although considered dioecious there is evidence from old oyster beds that sex ratio does not remain constant and indeed sex reversal takes place. The unfertilized eggs of *C. gigas* measure around 0·05 mm in diameter and an estimated 0·5–10·0 million are produced per adult. In general, after fertilization, the trochophore larva appears in 24 h and the veliger stage in 48 h with settlement at a size of 275–310 μm in approximately 14 days. In Hiroshima Bay the first spawnings take place at the end of May, when the water temperature is 18–20°C, and growing time till settlement is estimated at 16–20 days. Spawning tends to occur when spring tides carry warm offshore water of different salinity into the shallow water inshore culture areas. For late spawners, the development time, in temperatures of 24–26°C, is 12–14 days and swimming larvae of over 250 μm in shell height are considered near settlement stage (Ogasawara, 1972b). In the sea these larvae consume bacteria, flagellates and diatoms with diameters less than 10 μm, although in culture the Japanese have used monocultures of colourless, naked flagellates, such as *Monas* sp. (Imai and Hatanaka, 1949). Apart from temperature and feeding, salinity levels play an important role in larval development. *C. gigas*, being a littoral species, is said to develop more favourably in lowered salinities with uniform development and rapid growth being achieved at salinities of 23–28‰. Also vertical distribution of larvae in the sea can be correlated with particular haloclines (Ogasawara, 1972a) (see also p. 19).

## III. The Culture Areas

Oyster culture in Japan occurs in shallow coastal waters from the north of Hokkaido (Saroma Lake) to the south of Honshu and Kyushu (see Fig. 1). There are only two main spat collecting areas. Miyagi and Hiroshima supply the rest of the country with seed and also export considerable quantities to America, Korea and Europe (see Fig. 2). Miyagi is the principal seed producer with its colder winters providing an excellent seed hardening regime and 75% of its hardened seed is exported. The roles are reversed for ongrowing and production, Hiroshima producing more than five times the tonnage of Miyagi with oysters which are considered superior in quality. The hydrographic conditions prevailing in these two sea areas are discussed below.

### A.  *Hiroshima*

Hiroshima Bay is situated on the north coast of the Seto Inland sea, the whole area being sheltered by the large islands of Kyushu and Shikoku and numerous small islands within the Inland sea (see Fig. 3). Two branches of the warm Kurushio current influence the Seto Inland sea, flowing through the Kii and Bungo channels which have depths of less than 50 m.

In this almost enclosed area, seawater exchange is weak and on a reciprocal basis, the clockwise current flow produced by tidal rhythm. The sea area is about 400 km² with over 3500 ha used for culture inside the Bay. The area is sheltered by hills and mountains (the Chugoku range) and the shore varies from steep gradients which give 10 m depths (suitable for rafts) close inshore to moderate gradients which provide tidal exposed areas for racks, on bottoms composed of small well-rounded pebbles. The maximum depth in the Bay is 30 m.

The surface temperature in the Bay may reach 28°C in August with some stratification in the 15 m depths and up to 5°C difference between surface and bottom. In 10 m depths no stratification occurs between July and September. In the winter months (January–February) temperatures are around 10°C and rise quickly in the spring (see Fig. 4a). Therefore good growth conditions exist for most of the year. Hiroshima Bay, with the seven tributaries

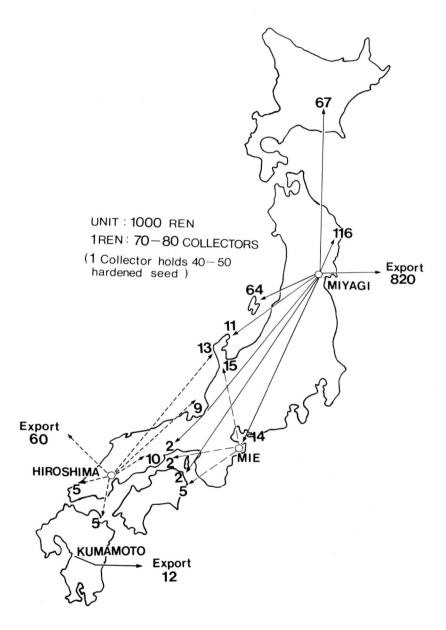

UNIT : 1000 REN
1 REN : 70−80 COLLECTORS
( 1 Collector holds 40−50 hardened seed )

FIG. 2. The production and distribution of oyster seed in Japan, expressed in 1000 ren units. Also the major exporting areas of Miyagi, Hiroshima and Kumamoto. (Ogasawara, 1972b.)

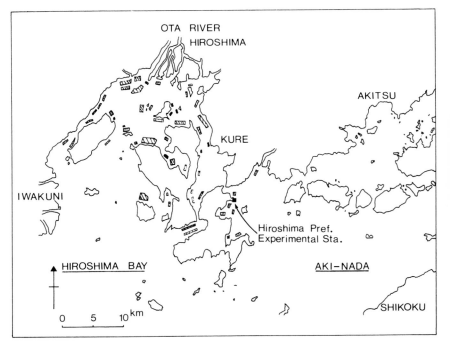

SOUTH WEST HONSHU

Oyster farming grounds in Hiroshima Bay

FIG. 3. The location of Hiroshima Bay in south-west Honshu showing the oyster culture grounds amounting to 3500 hectares.

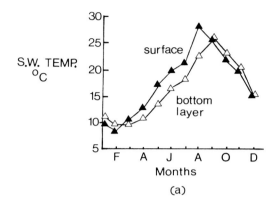

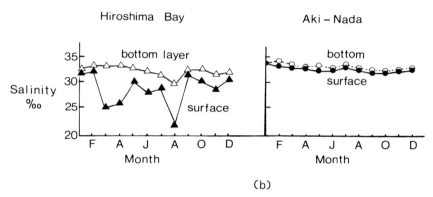

Fɪɢ. 4. (a) Hiroshima Bay annual mean monthly temperatures for surface and bottom layers (information from Hiroshima Prefecture Experimental Station). (b) Comparison of surface and bottom salinity inside Hiroshima Bay and Aki-nada to the east where oceanic salinities prevail (information from Hiroshima Prefecture Experimental Station).

of the Ota river flowing into it, acts as a catchment area for rainfall and in the rainy season (June) the inner Bay is covered with a 1 m layer of fresh water. Surface salinity ranges from 31‰ to less than 25‰, with the lower salinities occurring during the larval release and development months (June–August). Figure 4b compares the salinity range in Hiroshima Bay and in the Aki-nada area where oceanic salinities prevail and the exchange between the two areas has been calculated at 30 % per day. In the inner Bay, haloclines exist throughout spring and summer and are said to influence plankton movement.

Any discussion of the physical/chemical composition of the Hiroshima area inevitably involves pollution, since the chemical parameters now pre-

vailing in the Seto Inland sea are the result of the effects of industrial, agricultural and urban effluents. This is one of the most polluted regions of the
world in terms of area polluted, variety of pollutants and degree of pollution.
In addition, in Inner Hiroshima Bay, the organic deposits from the 9000
plus rafts have created a potentially eutrophic area. The Prefecture biologists
maintain that industrial pollution problems such as cadmium contamination
are no longer serious but that domestic detergents are responsible for the
increasing eutrophication (see pp. 36–38 for discussion of pollution problems).

## B.  *Miyagi*

The Miyagi coast with the main spat collecting and ongrowing areas between
Sendai and the Ojika Peninsula (Fig. 5) represents a very different coastal

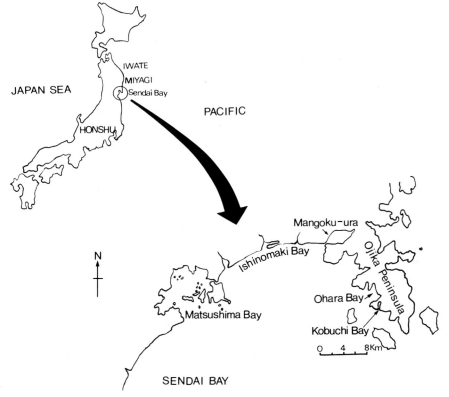

FIG. 5. The location of Sendai Bay and Ojika Peninsula on the Pacific coast showing
the major oyster grounds of Matsushima Bay, Mangoku-Ura and Kobuchi Bay.

situation. Here there is the direct influence of Pacific winds and currents affecting temperature and salinity and larval movements. The coastal currents of Sendai Bay are influenced by the Oyashio and Kuroshio currents which approach the Sanriku coast from north and south respectively and mix off-shore giving rise to Sendai Bay's peculiar water movements and larval distribution patterns (see p. 16). Although spat collection occurs in the ex-posed Sendai Bay, hardening and ongrowing takes place in sheltered bays such as Matsushima Bay, Mangoku-Ura Bay, Oginohama Bay, Ohara Bay and Kobuchi Bay. Over 2300 ha are used for culture.

These bays are a complete contrast to the completely oceanic conditions prevailing in Sendai Bay and different from each other. The well known Matsushima Bay, for example, is classified as partially eutrophic and has a 35·3 km² sea area only 2–3 m deep, connected to Sendai Bay by six water channels. Although summer water temperatures can reach > 28°C and organic pollution is high, the inside and outside of the Bay (another 35 km²) produce the nationally famous "nori" (*Porphyra* sp.) and hardened oysters. The waters of Mangoku-Ura of area 7·1 km² are classified as oligotrophic and are suitable only for the hardening of oysters on racks. Seawater exchange is poor and influx of nutrients low, resulting in mature oysters with low meat content known as "mizugaki" (water oysters). The bays of Ojika Peninsula are characterized by oceanic conditions and are heavily populated with rafts and long lines. The offshore water of Sendai Bay is 25°C in summer with salinity 31·5–33‰. The combination of these lower summer temperatures and higher salinities (Matsushima Bay's salinity is 27·0–30·8‰) are regarded as suitable for the growth of oyster larvae (Koganezawa, 1978). Figure 6 shows the average seasonal temperature and salinity variations inside Matsushima Bay compared to the oceanic conditions in Sendai Bay. Transparency ranges from 8 to 16 m in August in the outer Bay and down to 1 m at Matsushima Bay. The onshore winds prevailing at spat settlement time are less than 5 m/s. Matsushima Bay, like Hiroshima Bay, presents a remarkable example of how intensive culture in bays exists alongside industrial develop-ments. The Bay with its islands has a water area of only 70 km² and round the Bay exists a population of some 180 000 people and over 700 industrial establishments including food processing, metal, stone and cement producers, machinery manufacturing, and wood and furniture manufacturing.

Of these industries, however, more than 50% are food processing plants of marine products which pose a greater threat to the oyster and seaweed culturists, since organic pollution immediately affects growth and productivity as compared to the heavy metal and particulate pollutants of the other in-dustries which are gradually bio-accumulated. Organic effluent into Matsu-shima Bay has been estimated at 42 000 m³/day. However, pollution survey and treatment technology is well advanced in Japan, and Matsushima Bay

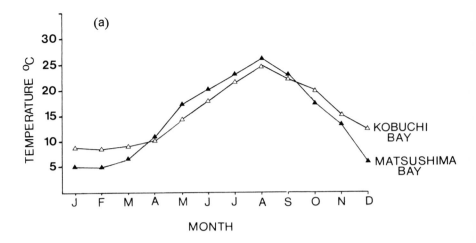

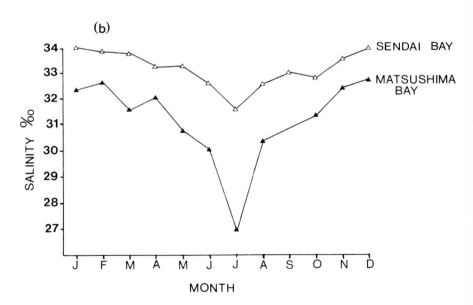

Fig. 6. (a) Average monthly temperatures inside Matsushima Bay compared to the oceanic temperatures in Kobuchi Bay (information from Ishinomaki Experimental Station). (b) Average monthly salinity levels inside Matsushima Bay compared to the oceanic salinity of Sendai Bay (information from Ishinomaki Experimental Station).

has been the subject of a large-scale dredging and water quality improvement scheme since the 1970s, aimed at maintaining and even increasing the culture productivity of the Bay (Watanabe *et al.*, 1972).

# IV. The Culture Technique

## A. *Spat Collection*

Spat collection and the production of seed oysters are so dependent on the combined characteristics of oyster stock and the coastal area, that settlement prediction and spat collecting strategy differ in each area. The maturation of gonads and the incidence of spawning of the parent stock influence the initial dispersal of the larvae, the spawning itself having been induced by the interacting effects of temperature and salinity. Further hydrographic, meteorological and topographical features then influence the distribution patterns and final setting of the spat. The main collection areas in Japan are Hiroshima and Miyagi and to some extent Mie and Kumamoto Prefectures where collection takes place on racks, rafts or long lines. The number of collectors (scallop shells) prepared every year for oyster spat collection amounts to around 500 million with some 300 million being set in Hiroshima Bay (Ogasawara, 1972b). Miyagi is well known for its seed hardening expertise and seed export, while Hiroshima seed is used mainly locally.

1. Hiroshima Bay

The currents in the Bay are clockwise and the main spat collecting areas used to be on racks offshore from the Yosijima area of the city. Consistent current patterns and south-west onshore winds in the Bay make plankton distribution easier to predict than in other areas. However, increasing land reclamation will probably mean that spat will be collected on rafts offshore in the future. The rack system produces more healthy, hardy spat whereas collection on rafts involves risks from certain predators and competitors (see p. 31). Details of rafts, and racks can be seen in Fig. 7. The Hiroshima raft, which is constructed with moso bamboo poles, has an area of approximately 200 m² and is buoyed by a total of 30 floats (three sets of ten floats along its length). The floats, each with a buoyancy of 0·3–0·4 tonne, are nowadays of styrofoam covered with replaceable polyethylene jackets. Hollow concrete buoys are still used in Hiroshima. The 2 m wire or "ren" holds 120 scallop shells separated by 1·5 cm bamboo spacers and is hung or simply laid over the racks and from 150 to 300 spat will settle on each scallop

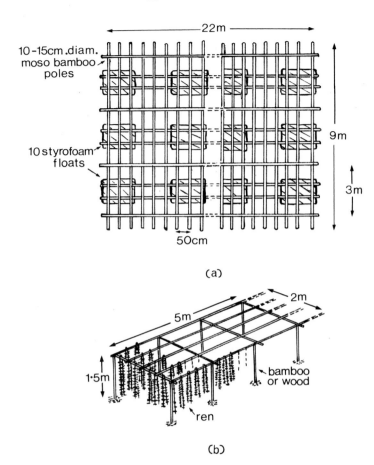

(a)

(b)

FIG. 7. (a) The Hiroshima raft formed of bamboo poles. Three main supporting pairs of poles lie lengthwise on top of the floats. Cross poles, 9 m long, are secured by wire perpendicular to these. Four strengthening poles, 3 m apart, are then tied lengthwise over the structure. (b) The Hiroshima rack from which collector ren are hung vertically for hardening intertidally (after Koganezawa, 1976).

shell. The preferred collector shell is the right valve of the southern scallop *Pecten albicans* Schroter ("itaya-gai") which is very durable, the surface microstructure being considered attractive to settling spat. However, these shells are scarce compared to the abundance of the northern scallop, *Patinopecten yessoensis* Jay ("hotate-gai") which is widely used but has the disadvantage of being large and collecting too many spat. Plastic manufacturers attracted by potential annual sales of hundreds of millions, are attempting to design a successful artificial collector based on the scallop shell. Some oyster shell is still used for collection.

The use of oyster extract on shell collectors to induce a gregarious settlement response has been tested by the Japanese in the field but the slight increase in settlement did not seem significant enough for commercial exploitation (Gillmor and Arakawa, 1979). Colour and texture of collectors has also been investigated by Ogasawara (1972b). His results indicate that rough surfaces collect more spat, and that black or dark surfaces are better than white or light surfaces.

Before setting the collectors there is an initial period of plankton forecasting and setting of test collectors. Plankton monitoring is undertaken for other larvae, such as barnacles, mussels and particularly *Hydroides norvegica* (Gunnerus) which can smother settled oyster spat. Euglenoid phytoplankton such as *Heterosygma* sp. are also checked since there is a correlation between this organism and *Hydroides*, both being associated with increasing eutrophication (see p. 36). Sea squirts are also a serious pest but their peak settlement is impossible to forecast. The first spawning of oysters in Hiroshima Bay is around the end of June/beginning of July. The water temperature is then normally 18–20°C and the veliger stages take 16–20 days till settlement

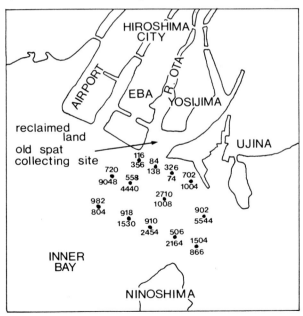

FIG. 8. The distribution of oyster larvae in inner Hiroshima Bay at the spat collecting sites. Numerators and denominators indicate the larval numbers/100 ml at flood and ebb tide respectively. (After Ogasawara, 1972a.)

or 12–14 days at higher temperatures of 24–26°C. Spat size after settlement is about 275–310 μm and swimming larvae > 250 μm are considered to be about to settle. Larvae from vertical plankton hauls are measured into three categories viz: small larvae < 150 μm; medium larvae > 150 μm– < 250 μm; large settlement size larvae > 250 μm–300 μm diameter (foot and eye spots visible). Larval distribution in the Bay is very heterogeneous with dense masses of larvae being moved by tidal rhythm and also concentrated by haloclines and thermoclines. Fig. 8 shows the horizontal distribution of oyster larvae in the inner Bay at flood and ebb tide. No correlation can be observed, however, between tidal rhythm and change in larval numbers (Ogasawara, 1972a). The traditional spat collecting areas are on racks in-shore at HLWN level (highest low water level of neaps) although plankton estimates each year indicate that more spat could be collected offshore.

Larval numbers greater than 200/100 l are considered adequate for a good setting of around 200 spat/collector shell (see also Wisely *et al.*, 1978). Settlement prediction is based on the occurrence of larvae > 250 μm and on appearance of spat on test collectors which are sampled every two days. The Prefecture laboratory (Nansei Regional Fisheries Research Laboratory) issue spat fall data and the oyster culturists decide when to immerse their collectors. The period of 3–5 days after immersion is thought optimal for spatfall. Spat collection proceeds from the beginning of July and through August with double or triple wires (ren) being set out, their spacing being dependent on the predicted density of settlement. Spat which settle on racks stay there for a hardening period which can vary from one to nine months of a normal two-year culture period prior to final growth on rafts.

## 2. Miyagi

Prediction of spat settlement and collection of spat on this coastline is more difficult than in Hiroshima Bay, because the Sendai Bay and Ojika Peninsula face the Pacific and the current and wind patterns are of much greater dimension. The confluence of the cold Oyashio current and the warm Kuroshio and Tsugaru currents, and their salinity and temperature effects, move plankton inwards or out of the Bay. Larval groups from different sources are mixing in the Bay and are generally moved from north to south. The larvae released into Sendai Bay are thought to come from three main parent stock viz: Matsushima Bay oysters which spawn early and in abundance; the Mangoku-Ura oysters which also spawn early but in small quantities; and the Ojika Peninsula group which spawn late in large quantities. Spawning originates from natural oysters and one-year-old and two-year-old cultured oysters, although most occurs in oysters which are finishing their first year of culture. The Matsushima Bay oysters which spawn first in mid-

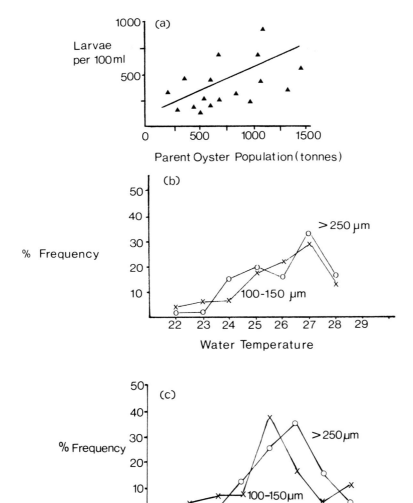

FIG. 9. (a) Relationship between the parent oyster stock in Matsushima Bay and the number of oyster larvae/100 ml after spawning (data for 1953–1969, Ishinomaki Experimental Station). (b) Frequency of occurrence of pre-settlement larval groups related to increasing seawater temperatures in Matsushima Bay (data from Ishinomaki Experimental Station). (c) Frequency of occurrence of pre-settlement larval groups related to increasing salinity in Matsushima Bay (data from Ishinomaki Experimental Station).

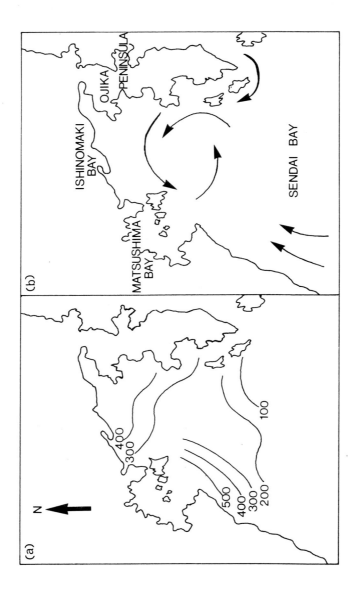

FIG. 10. (a) The larval distribution (larvae/100 ml sea water) in Sendai Bay before settlement and (b) the prevailing current systems in Sendai Bay which concentrate the larvae (after Koganezawa and Ishida, 1973).

July (three spawnings) at the tidal changeover from spring to neaps contribute most to the collection of spat in Sendai Bay (Koganezawa and Goto, 1972).

Figure 9a plots the average number of larvae 100 l/100 ml in relation to the quantity of parent oysters in Matsushima Bay. About 200 larvae $> 250$ μm is favourable for a settlement of 50–200 spat/collector shell. As mentioned previously, temperature and salinity conditions affect the spawning and larval development and Fig. 9b,c shows the frequency of appearance of the different larval size groups in relation to prevailing temperature and salinity conditions in Matsushima Bay. Larval numbers peak when temperatures reach 27°C and the specific gravity of sea water is 21–23 (31·0–32·5‰ salinity). In all the inner bays of Sendai Bay, as larvae are being released, large quantities of colourless flagellates (averaging 6000/ml) provide good feeding for the oyster larvae. These flagellates are thriving on decomposing *Zostera* ("amamo") which grows abundantly in all the bays and withers about oyster spawning time (Koganezawa, 1972). Some 24 000 tonnes (dry weight) of blooming *Zostera* have been estimated in Matsushima Bay alone.

The coastal current systems and wind patterns initially accumulate larvae in Sendai Bay but the final distribution patterns are governed by the eddies which are formed by the effect of the coastal geographical features on tidal streaming. Figure 10 shows the typical larval distribution pattern in the Bay. The characteristics of the coastal currents also affect growth and development of the larvae, combinations of lower temperatures (24–26°C) and higher salinities (28·2–32·0‰) offshore being optimal for larval growth as compared to the inner Bay conditions (27–28°C and 25·5–29·5‰) where spawning begins (Koganezawa and Ishida, 1973). In Sendai Bay larval numbers of 200/100 l sea water are considered necessary for good spat fall and usually the larval concentrations are up to 50 times this level. In early August there is a settlement of around 150–200 spat/collector which are reduced to 40 or 50 after hardening.

Seed are hardened from August to September for local use or from August to March for export. Most cooperatives set their scallop shell collectors offshore on long lines which are parallel with the coast and at right angles to current flow. Figure 11 shows a typical 90 m Miyagi long line with 1500 2 m strings of collectors with each collector holding 70–80 shells.

## B. *Hardening*

The modern practice of hardening arose when the Japanese culturists began hanging culture on rafts in the 1920s. Here the oysters were constantly immersed as opposed to the intermittent exposure experienced by natural oysters and bottom culture oysters subjected to tides. Mass mortalities

90m- LONG LINE
1500 - REN
  25 - BUOYS(final no.)

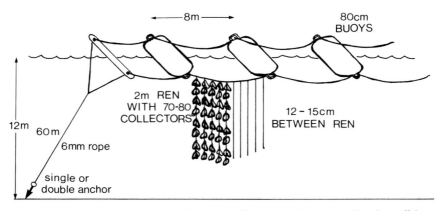

8m

80cm
BUOYS

2m REN
WITH 70-80
COLLECTORS

12 - 15cm
BETWEEN REN

12m
60m
6mm rope

single or
double anchor

Fig. 11. The Miyagi double long line with collector ren for spat collection offshore. Buoy numbers are increased from an initial ten to 25 as the installation sinks due to settlement and fouling.

amounting to 50–90 % occurred almost annually in those areas where raft culture had been introduced, and from 1945 severe mortalities of 20–100 % occurred on rafts in the Hiroshima area. After the war, seed were being produced for an export industry which, under American direction, would jointly revive Japan's oyster industry and provide seed oysters for the West (Cahn, 1950). Seed oysters for export were often stocked on racks for convenient holding before transport and it became evident that those export seed and oysters whose growth had been stunted survived better. It was also noticed that the natural oysters suffered no such mass mortalities. Even in the bottom sowing culture methods of the previous centuries the oysters on the shore had been raked around to increase exposure and thus promote "harder" shells, a concept which was to be revived and modified for hanging culture hundreds of years later. Two lines of research were carried out, involving growing oysters rapidly on rafts after settlement on racks, or exposing spat on racks for an extended period before raft culture. These two lines of research eventually gave rise to the one-year and two-year culture systems in the Hiroshima area. The exact cause of these mass mortalities was never clarified, although the general opinion was that with the greatly accelerated growth and gonad development of raft hanging oysters, their spawning became impeded owing to particular seasonal combinations of high water temperature and high salinity, leading to death of the spawning stock (for discussion of abnormal mass death, see p. 34).

The results of comparative trials in different areas with seed oysters hardened to various degrees and seed put immediately into hanging culture have led to some understanding of the mechanisms of hardening and why survival is enhanced. A summary of these findings is given below (Ogasawara *et al.*, 1962):

*Growth*
(a) Seed undergoing hardening grow at half the rate of raft culture seed.
(b) Survival during hardening period was twice that of those seed on the rafts.
(c) Hardened seed oysters of size 29 mm (10–11 months of hardening) and raft oysters of the same age at 81 mm, when ongrown for a further three months resulted in hardened oysters 80 mm in height and raft oysters only a few millimetres bigger than their original size. Survival during this period was around 78 % for hardened and 57 % for raft seed. This is thought to be the result of the quick restoration of connective tissue and fattening after spawning in the hardened oysters as compared to the prolonged spawning period of the raft oysters.
(d) Time of transfer of the rack hardened seed to rafts indicates that the earlier the transfer, i.e. shorter hardening period, the slower the growth during that spawning season. If hardened oysters are transferred to rafts after a short hardening period, they spawn in a similar fashion to raft oysters.

*Gonads*
(e) The energy expended during spawning was far less in the hardened seed (measured by pre- and post-spawning water and glycogen content of the meat) and the area occupied by the gonad in the raft seed was twice that of the hardened seed (69 % in cross-sections).
(f) Development of the gonad was later in the hardened seed and over a short period, and only small numbers of gametes were released. In the raft culture seed, the gonad developed earlier and gradually with a large release of gametes followed by the "water" oyster condition. However, in the hardened seed after spawning, the gonad was immediately absorbed and the "water" condition seldom existed.
(g) There are a greater proportion of males and single sex individuals in the hardened seed.
(h) Individuals with abnormal multinuclear eggs did not occur in hardened seed.

*Filtration Rate*
(j) The reaction of hardened oysters to temperature or salinity changes is a marked increase in the rate of ciliary movement, perhaps indicating greater adaptability to varying environmental conditions. This rate

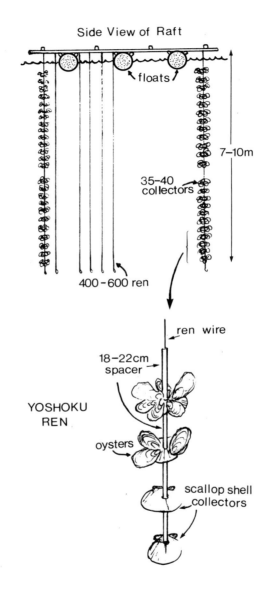

Side View of Raft

floats

7–10m

35–40
collectors

400–600 ren

ren wire

18–22cm
spacer

YOSHOKU
REN

oysters

scallop shell
collectors

Fig. 12. Yoshoku ren for ongrowing to harvest. There are 35–40 collector shells with seed oysters per ren and 400–600 ren/raft (3 ren/m²).

reached a maximum at experimental temperatures around 32°C and salinities around 15–17‰.

These experimental findings led to the widespread use of hardened seed in the Japanese oyster industry although one-year culture made up 70 % of Hiroshima's total production for many years. Two-year culture consisting of one year of hardening led to economies on raft holding and improved survival and meat quality of oysters.

## C.   Ongrowing in Hiroshima

After spat collection from the end of July, the strings of collectors are left on intertidal racks for hardening. In previous years one-year culture was possible in Hiroshima Bay, the settled spat on the scallop shells being restrung to form "yoshoku" ren (see Fig. 12) in August–November after a short hardening period. These oysters transferred to rafts could then be harvested early the next year, from January onwards, at medium size category. The larger size categories were obtained from two-year culture in which oysters were restrung on "yoshoku" wires in spring of the following year after nine months of hardening and transferred to rafts where they would grow to produce meats of > 30 g to be harvested from October onwards. These growth rates of 6–12 months of ongrowing to commercial size no longer exist today because of water quality deterioration (see p. 36) over the last 15 years. Oysters must now be ongrown for 2–3 years to reach commercial sizes with meat weights around 15–20 g. In recent years, therefore, all ongrowing has been an extended version of two-year culture. One-year culture is still occasionally referred to in the literature however (Korringa, 1976; Honma, 1980). Figure 13a shows the original one-year and two-year culture schemes and Fig. 13b shows the growth rates achieved in each scheme. One-year culture is no longer possible since the oysters cannot reach the minimum market size by early the following year, and their survival is poor in that year because of the very short hardening period.

Hardening technology is better developed in Miyagi (see p. 19) where oyster seed are prepared for export. In Hiroshima, hardening is utilized mainly to kill off harmful competitors and predators (see p. 32) and the racks are positioned just below highest low water level of neaps (same for collection) where spat are daily exposed for 3–12 h. Survival of hardened seed in the 2–3 year culture period is 70–90 % compared to 20 % with no hardening. During the hardening period, the 150–300 settled spat (per collector) are reduced to 40–50 viable oysters for ongrowing which will finally be reduced to 20–30 harvestable oysters. After hardening on racks the collectors are restrung on 10 m wire at 21 cm intervals to give 35–40 collectors per "yoshoku"

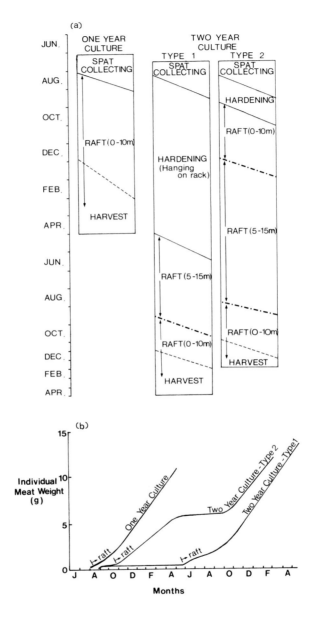

FIG. 13. (a) Plan of one-year and two-year culture schemes in Hiroshima Bay (after Fujiya, 1970). (b) Growth rates in one-year and two-year culture schemes (data from Hiroshima Prefecture Experimental Station).

ren (see Fig. 12). The spacers between the shell collectors are preferably bamboo, with PVC-coated bamboo and PVC tubing of recent use, although the latter are considered inferior until "aged" in the sea. Plastic spacers and also collectors tend to expand in the summer and cause oysters to fall off. In June these ren are then hung from traditionally designed rafts made from Moso bamboo which are usually 20 m × 10 m (see Figs 7a and 14). There are up to 600 ren/raft which can be hung by five people in 2–3 h. The 600 ren therefore carry around 600–800 oysters/ren with a total 450 000 oysters growing to harvest size. Only about one-half these oysters reach commercial size ($\geqslant$15 g meats) at the beginning of the season and therefore about ten oysters/collector are harvested. This gives a theoretical yield of around 4·5–6 kg of meat/ren and 2·5–3·5 tonnes/raft. Growth under the rafts is not uniform with those oysters in the centre growing slower than those on the outside which directly face the currents. Current measurements have shown that a current speed of 8–10 cm/s is reduced to 2–3 cm/s at the centre of the raft and oysters are therefore harvested in sequence from outside inwards. Groups of 4–5 rafts are arranged in line with water currents or at right angles. When parallel with water currents, growth tends to be more even throughout the raft. At right angles to water flow, growth is reduced at the centre of the raft but faeces are carried far off to one side since the raft movement can be 60 m on ebb and flow. The recommended distance between groups of rafts is now 300–500 m, and if raft positions are changed between harvests, oyster growth and quality is better and the ground can recover from the raft deposits (see p. 37). The number of rafts in Hiroshima Bay

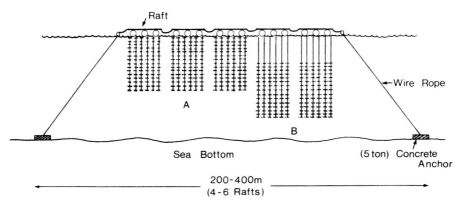

FIG. 14. A group of Hiroshima rafts moored for ongrowing. The rafts in group A have oysters growing from 0 to 10 m, and in group B from 5 to 15 m in order to utilize the water column efficiently (information from Hiroshima Prefecture Experimental Station).

utilizing an area of 35 km² had risen to around 11 000 by the 1970s and the number of "yoshoku" ren had risen from the optimum of 400/raft to 550–600/raft as the growers attempted to exploit available raft space. Through recommendations from the Prefecture laboratories and the evidence of diminishing returns, the number of rafts was reduced to < 10 000 by the middle 1970s and currently there are around 9000 rafts (see Fig. 15a). The growers are organized into about 600 groups which manage about 15–20 rafts each. The growers are now being advised to reduce the number of ren on rafts to 400 to improve meat yields which have fallen from 3 to 4 tonnes meats/raft to an average 2·0–2·5 tonnes/raft in the last 20 years (see Fig. 15b). 1979 and 1980 production, however, indicated yields in excess of 3·0 tonnes/ raft. The servicing of rafts, and the lifting and transferring of yoshoku ren, is achieved with the characteristic tall masts and power winches of the Hiroshima oyster work boats. The 15-ton boats with 8 m high masts are powered by 60 hp engines. Smaller boats with 3–5 hp engines are used for transferring collector ren from racks to rafts.

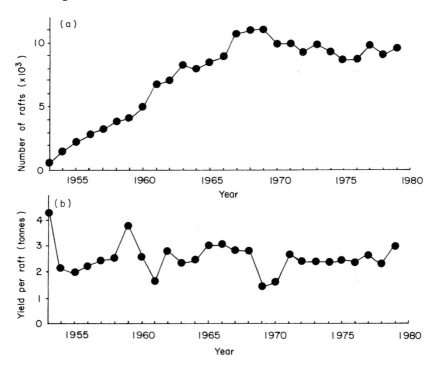

FIG. 15. (a) Trends in number of raft installations in the last 25 years (data from Hiroshima Prefecture Experimental Station). (b) Annual trends in raft yields in Hiroshima Bay (shucked meats).

Site selection is very well understood in Japan and within the Hiroshima area different sites give different rates of shell growth or meat weight increase, or time and duration of spawning; even within one locality these parameters can vary with the depth at which the oysters are suspended. During ongrowing, oysters can be transferred from rafts in one locality to another with particular results, for example, oysters grown in less favourable sites when moved to good growth sites will show improved rates of growth and fattening and raft yields which are better than those for oysters which were grown there initially. In general, growth and fattening can be accelerated by transferring oysters at different growth stages to the appropriate culture ground. Different culture sites in the Hiroshima area are therefore recognized for promoting shell growth or fattening or accelerating spawning. By transferring oysters and regulating hardening, the culturist can take advantage of these site properties and promote fattening or spawning when necessary.

The problems of overcrowding in the raft culture areas can also be managed by careful attention to the fertility of the different areas. The effect of over-population on growth and fattening of oysters depends on the concentration of available plankton. In areas where the micro-suspensoid (measure of micro-organism concentration) is $> 1.5$ mg/l, growth and fattening can proceed at high rates regardless of the number of oysters per ren. Where the concentration of micro-organisms is less than $0.8$ mg/l, growth is invariably poor, again regardless of raft stocking density. At values between $0.8$ and $1.5$ mg/l, however, there is a direct relationship between raft stocking density and growth and fattening (Ogasawara et al., 1972). It has also been noticed that oysters growing in the more fertile areas are less adversely affected by fouling organisms. The above knowledge has been used to produce and market oysters on schedule and stabilize the industry, although overstocking generally exists and recently deteriorating water quality has significantly controlled production.

## D. Ongrowing in Miyagi

Oyster culture on the north-east Pacific coast of Japan differs principally from Hiroshima in that as well as rafts, long lines are used for hanging culture off the coast. Whereas Hiroshima has long been recognized as the major producer of fast growing oysters of prime size and quality, Miyagi oysters were considered inferior in taste and the culture techniques concentrated on spat collection and hardening of spat (much of it for export). Miyagi growers, however, would contend that nowadays with the water quality problems in the Hiroshima area, the Miyagi oyster growth is comparatively fast in the colder waters, the oysters are resistant and have high survival rates, and quality is very constant from year to year. The main problems facing the Miyagi growers are first the progress of seaweed culture (mainly "nori") which competes for long line space and is consistently profitable since Miyagi "nori" is of premier grade. Secondly, production is falling behind demand and thirdly, the phenomenon of mass abnormal death has occurred since 1961, but has been interpreted as due to combinations of adverse environmental conditions (temperature, salinity, oxygen deficiency etc) in particular years (Kanno et al., 1965). Certainly the move to more offshore long line spat collection and culturing should alleviate these problems.

The production of hardened seed from Miyagi is considerable, formerly averaging 3·25 million ren a year, i.e. around 10 billion 10 mm seed. Of this, 970 000 ren were used in Japan with around 1·5 million ren exported to America and Europe and the rest to countries such as Korea and Australia. In the 1970s, seed production averaged over 1 million ren (see Fig. 2) and latterly has remained at that level. The hardening method in Miyagi involves taking up the collectors with their settled spat (around 200/collector) in September and laying them horizontally across the racks. In Hiroshima the ren for hardening are hung vertically as for collection, which takes up less space, but horizontal laying on the racks results in more uniform consistent hardening, since all the spat from top to bottom of the ren receive equal exposure between tides. For domestic use, the strings of collectors can be sold by October, after one month of hardening (for further hardening). Of this domestic seed 20% is used in Miyagi, 50% is sold to neighbouring Iwate and about 10% is sent north to Hokkaido, with smaller amounts to Shizuoka in the south. For export, the spat are hardened for 5–7 months until 10 mm (by December) or 15 mm (by March). During hardening, the initial settlement of 200 spat is reduced to 40–50 seed/collector which will result in 20–30 ongrowing oysters/collector shell, taking two years to commercial size with about 50% harvested. From April of the first spawning season to January harvesting in the next year, survival will vary from 40 to 80% according to the duration of the hardening. Shells can reach 80 mm in height by June. For ongrowing, 8–10 m strings, holding approximately 500 ongrowing oysters, are hung from double 60 m long lines, buoyed at the surface, usually in shallow depths of less than 12 m (see Fig. 11). One such long line holds 280 ren at 30 cm intervals with approximately 60 000–70 000 harvestable oysters/line. This gives a yield of around 3·5 kg of meats/ren and about 0·95 tonnes of meats/long line at harvest time (Koganezawa, 1976). Long lines are used where typhoons would damage rafts which cannot withstand wave heights greater than 1 m. The Miyagi raft differs from the Hiroshima raft in being made of cedar spars and is much smaller measuring around 9·1 m × 4·5 m (see Fig. 16). About 150 5–9 m ren can be hung from these structures, which are only suitable for sheltered areas

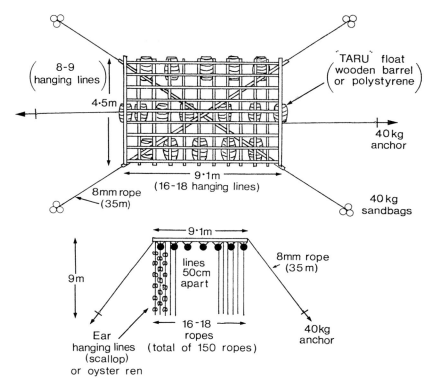

FIG. 16. The Miyagi raft constructed of cedar spars. Area 40 m².

inside bays, the rafts usually being anchored singly. The yield/ren is about 3·25 kg, some 0·5 tonne per raft. The smaller Miyagi raft is only slightly cheaper to make than the twice-as-large Hiroshima raft; but is cheaper than the long line (see p. 31). The holding capacities are considerably different, however, with the Miyagi raft capable of holding around 90 000–100 000 oysters growing to commercial size, the Hiroshima raft 420 000 and the Miyagi double long line 250 000 oysters. Long lines for oyster culture are being increasingly used in Japan, now accounting for 50% of total raft and long line installations as compared with 20% ten years ago.

## E.  Culture Costs

Production and market values are dealt with on pp. 38–52 but presented here are some details on equipment costs of rafts and long lines (excluding boat costs) and recent values of seed oyster and shucked meats to the co-operative.

## 1. Hiroshima

The Hiroshima raft of 200 m² has a life of over three years if maintained properly and its cost for bamboo and wiring would be 200 000 yen (£400) at 1979 prices. Rental charges would add an extra 8000 yen/raft. The costs of the hanging components are as follows: scallop shells about 1·5 yen a piece and 50 % can be used for a second season; a scallop shell with attached spat about 5–6 yen; a single ren wire of 2 m less than 100 yen. A single ren of collectors with spat would therefore be valued at 600–800 yen and a single raft would carry 360 000 yen worth of collectors plus spat, with the spat settlement valued at around 250 000 yen/raft. At harvest time a single raft would carry about 420 000 oysters, 50 % of which could be harvested, yielding some 3000 kg of meat in a good year, with a market value of over 2·5 million yen (£5000). Therefore the average family group with around 20 rafts would harvest say 50 000–60 000 kg of meat in a good season worth approximately 40–50 million yen at around 30 % profit. Production costs are currently around 620 yen/kg of meats and market value 1100–1250 yen/kg, giving profits of 70–90 % on fresh meats although about 35 % of production will be processed at prices less than half those for fresh meats and an average price might be 850 yen/kg (see p. 49). Wages vary for shucking from 2000 yen/ day for a 7-hour day to double this for skilled shuckers (usually women) who can reportedly shuck up to 3500 oysters/day, producing a surprising 50 kg of meat. Most shuckers probably produce about 25 kg of meat/day.

## 2. Miyagi

In Sendai Bay, long lines are used for spat collection and ongrowing and the costs related to these. A Miyagi double long line of 60–90 m would cost 250 000 yen (£500) including buoys and anchorage. The 80 cm wide sausage buoys cost up to 7000 yen each and there may be 25/long line for spat collecting or up to 30 buoys for ongrowing. One ren of collectors consisting of 80 shells costs from 100 to 120 yen and a 90 m double long line would hold up to 1500 of these ren of value 150 000 yen. Thus the long line plus collectors are valued at around 400 000 yen and have a life of five years (replacing shells). After spat settlement the ren plus spat are valued at 500–600 yen and the long line holds an average 825 000 yen worth of ren with the spat settlement/long line worth about 660 000 yen (see p. 20). Each co-operative member would operate four such lines.

For ongrowing in Ishinomaki the ren are stocked on 60 m double long lines at 30 cm centres (15 cm centres for spat collecting) giving 200 10 m ren/long line. A long-line ren in Ishinomaki yields about 5 kg of meat (20 g each meat) and the whole production in co-operatives like Ishinomaki is of fresh shucked meat which has a current peak market value of 25 000 yen per 20 kg pack or 1250 yen/kg (1979–1980 price). Thus an Ishinomaki double long line produces an average 2·5 million yen worth of oyster meats (2·0 tonnes). Shucking wages and production rates are similar to those of Hiroshima.

## V. Culture Problems

Oysters in intensive culture can be affected in many different ways by predators, competitors, parasites and changes in water quality which can act chronically on growth and fattening, or acutely, resulting in large scale mortality. The oyster culturist can combat predators and fouling by careful scheduling of his operations and to some extent can control water quality around his rafts by stocking sensibly, although he is at the mercy of the regional industrial contamination of the local sea area. The recent pollution problem causes additional stress on top of the occasional outbreaks of fouling and predatory organisms and is one problem against which the culturist cannot safeguard.

### A. *Predators*

One of the most well documented and serious oyster afflictions is due to the predaceous turbellarian *Stylochus iijimai* (Yeri *et* Kaburaki) which can affect 40–60% of the oysters in Hiroshima Bay. It occurs offshore where high salinity prevails and if this condition is combined with high temperatures and low rainfall, there is a serious risk from this flatworm. *Stylochus* penetrates the shell of maturing oysters in hanging culture and consumes the flesh by secreting proteolytic enzymes. Another oyster-feeding turbellarian *Pseudo-stylochus ostreophagus* (Hyman) attacks specifically young oysters and at seasonal temperatures of 15–20°C swarming of this predator takes place, and 30–40% of all seed oysters on racks can be infected. The oysters are either killed by the drilling or smothered by the excretion of mucus from the predator. The periods of greatest damage coincide with the seed hardening periods and as a precaution, seed oysters should be hardened at the highest tidal level. When maturing oysters have been attacked they are subjected to freshwater treatment (30 min immersion) or hypersaline treatment (30 s immersion in 20% salt solution). Another group of serious predators are the oyster drills, such as *Ocenebra japonica* (Dunker), *Rapana thomasiana* (Crosse) and *Thais clavigera* (Kuster). They inhabit sandy, muddy and reef areas respectively and are remarkably euryhaline. The most serious pest is

*Ocenebra* which spawns twice a year, in spring and autumn, with the trochophore larvae developing inside the ootheca intertidally and emerging as
young snails at the high tide level. It therefore lives at the same depths as the
seed oysters, and causes most damage during the seed hardening period of
August–December (oysters on rafts are usually unaffected by drills). This
predator can be controlled with baited traps and burning, which is preferred
from an ecological standpoint to the carbamate chemical treatment where
seed oysters are immersed in a 30% solution of 1-naphthyl-$N$-methyl-
carbamate which is reputably lethal for gastropods but does not harm
bivalves (Koganezawa, 1978). This chemical, mixed with sand, can also be
spread around the culture grounds to prevent oyster drills moving in. Other
mollusc predators such as starfish and crabs are not a great threat to oysters
on racks or rafts.

## B. *Competitors*

The settling stages of other marine larvae and algal species can compete with
the oyster spat for space and food. The main such competitors were formerly
mussels, sea squirts and barnacles. In 1969, however, in Hiroshima there
occurred an unprecedented outbreak of the serpulid tube worm *Hydroides
norvegica* (Gunnerus) which seriously damaged some 6000 rafts of oysters
at a time when the raft population had risen to the high level of 11 000 rafts
and 40% of the total harvest with a market value of some 3 billion yen was
lost (Arakawa, 1971). Another serious outbreak occurred in 1970 with a
peak settlement at 5–6 m of more than 300 young worms/100 cm² on the
spat collectors. This serpulid worm now occurs annually at average concentrations of around 20–30 worms/cm² and its presence indicates the
chronic eutrophication existing in the Bay with the worm outbreaks related
to the presence of euglenoid phytoplankton. The major settlement occurs
from the middle of September to the middle of October, affecting both seed
and maturing oysters and smothering them and competing for food (Arakawa,
1973a). In the major outbreaks in 1969 and 1970 some areas had up to 180 g
(wet weight) of worms/collector competing with a similar biomass of oyster
seed. Each year the presence of *Hydroides* in the plankton is forecast on
a short term basis and collectors can be set with respect to time and depth
to avoid peak settlement (Arakawa and Kubota, 1973).

Another fouling organism which acts in a similar way to *Hydroides* is the
blue mussel *Mytilus edulis galloprovincialis* (Lamarck). Fouling by mussels
was always an endemic problem in Hiroshima Bay and they were the major
competitors for food before the outbreaks of *Hydroides* in the late 1960s.
There was, however, an unprecedented outbreak of mussels in 1973 in one
area of the Bay, around Ondo island where damage to 60% of Hiroshima's

production occurred (Arakawa, 1974). Maximum settlement of mussels occurs around 1 m below the surface down to 5 m. Mortalities in oysters can range from 20 to 60% in the top 5 m with 20–30 mussels settled/oyster falling to less than ten/oyster at 5 m. Therefore the heavy mussel fouling can be avoided by hanging oysters below 5 m, at which depths meat yield is also improved. For example, hardened seed held at 5–10 m depth can average 2–7 g of meat (wet weight)/oyster as compared to 3–4 g at less than 1 m. There are several treatments for fouling mussels of which the simplest and most practical are exposure to air and brush cleaning, or hydro jet cleaning (28 kg/cm² pressure) for moderate fouling. On a hot summer day, 2–3 h of exposure to air will cause the internal temperature of mussels to reach high figures of 38–40°C, when death occurs. The oysters survive, having a higher lethal temperature threshold of over 44°C (Arakawa, 1973b). With heavy settlement, petroleum burners have been used which burn through the thinner mussel shells and are also effective against all other fouling. Another effective treatment is hot/cold water dipping in which the fouled oysters are immersed in hot water (70°C) for 15 s, which can enter the byssal gape of mussels but does not affect oysters. They are then returned to ambient sea water. This treatment is very suitable for spat which might be damaged by the other treatments (see Arakawa, 1973b for further methods).

Other less serious competitors for space and food include barnacles whose peak settlement can be avoided by attention to plankton forecasts, and by time and depth of placement of the collectors. In the last ten years, there have been occasional heavy settlements of the hydrozoan *Tubularia* sp. which is associated with the high nitrate levels in Hiroshima Bay. The boring worm *Polydora* sp. is also present in the Bay but does not inflict serious damage on raft hanging oysters. Seaweed fouling does not affect the seed on the exposed racks and is not a serious problem for oysters on rafts.

## C. *Parasites*

Although some of the species mentioned earlier are sometimes classified as parasitic, they are considered here to be essentially predatory or fouling organisms competing for food (Arakawa *et al.*, 1977a). This section deals with those parasitic infections which invade cells and can lead to pathological changes. Among the parasitic organisms causing disease in oysters are micro-organisms such as viruses, bacteria, protozoa and fungi and the parasitic flatworms, trematodes and cestodes (Arakawa *et al.*, 1976).

In Japan, serious pathological changes in cultured oysters such as enteritis and multiple abscesses are sometimes encountered during the spawning season. These are associated with bacterial infection and amoebocytosis.

However, the mass deaths which sometimes accompany these conditions (see p. 35) are not considered by Japanese workers to be directly due to bacterial or amoeboid infections. These infections are considered secondary, following physiological and metabolic disorders during spawning brought on by adverse temperatures and eutrophication at spawning time. Nevertheless, there is an abnormal occurrence in the Hiroshima area outside of spawning time (late September onwards) when egg masses appear in the form of neoplasia (multiple tumours). Histological examination of such gonads took place in 1974 (Matsuzato *et al.*, 1977) when one or more enigmatic bodies resembling nuclei were found in the cytoplasm of ova in the abnormal oysters. These bodies, over 25 μm in diameter, had a thin membrane which stained with eosin and were PAS-positive, and they were considered to be parasites. They occurred in up to 12% of oysters in oligotrophic areas of high salinity and were present in oysters ranging from 40 to 116 mm at depths down to 9 m. This phenomenon of abnormal egg masses was first described in 1933, soon after the introduction of hanging culture, and is now the subject of electron microscope studies. It is not thought to be associated with mass deaths.

### D.  *Mass Mortality*

Shortly after the introduction of raft culture in 1927 mortalities of raft hanging oysters amounting to 50–90% occurred in the Miura Peninsula area at the mouth of Tokyo Bay. The only environmental factors correlated with the deaths were high water temperatures and high salinity during the spawning season. After the war in 1945, raft culture techniques spread. All over western Honshu and in Okayama, Tottori and Shimane Prefectures, and particularly in Hiroshima Bay, in the 1950s and 1960s, oyster farms experienced periodic mass mortalities ranging from 20 to 100%. A mass mortality is defined as the death of 50–60% of the oysters, with >60% mortality resulting in commercial failure. During this period, similar mortalities were occurring on the Pacific coast in Miyagi Prefecture, leading to intensive investigations of the affected environments and development of countermeasures (Kanno *et al.*, 1965; Imai *et al.*, 1965). Common to all of these outbreaks since 1927 were the following features (Koganezawa, 1975).

(1) Only adult oysters (1 year old) in their spawning season were affected.
(2) Highest mortalities were observed in the faster growing oysters.
(3) Mortalities occurred gradually when water temperatures rose to 21 or 22°C and increased relative to the rising water temperature or salinity. However, maximum mortality did not coincide with maximum temperature or salinity.
(4) Mortality was not related to depth or distance from shore.

(5) In Hiroshima Bay, the hydrographic conditions were not significantly different in the high mortality year than in other years.

Ogasawara *et al.* (1962) considered the above factors and those others which have been considered to cause death (singly or in combination) in oysters in other countries as well as Japan. They formulated the following categories.

(1) Physiological impairment due to abnormal low water temperature.
(2) Physiological impairment due to abnormal low salinity.
(3) Physiological impairment due to abnormal high water temperature and/or high salinity.
(4) Insufficient solar penetration (due to cloudy/rainy weather), resulting in shortage of food.
(5) Unfavourable water quality (artificial or natural cause).
(6) Parasites (include *Polydora*, *Hydroides*, Turbellaria).
(7) "Simple" disease.
(8) Pathogenic micro-organisms.

The Japanese workers realized that the cause of these mass deaths, whether it was of a chemical, physical, or biological nature or a combination of these, was such that it was inseparable from the oyster farm and its particular environment. They sensibly admitted that preventive measures aimed at "improving" the environment were very difficult to achieve when the clues to the cause of these disasters were so indefinite. They decided that artificial counter-measures would be insignificant and that the cause of these mortalities would always be present. They then took the bold step of deciding to modify their culture methods in the hope of improving the condition of the oysters and imparting some resistance to mass mortality. As well as investigating the mortalities they closely analysed those oysters which had not suffered mortality. The following facts emerged:

(1) Those oysters whose spawning was delayed (slower growing ones) did not suffer mortalities.
(2) The natural oyster beds were unaffected.
(3) The spat which settled during the mass mortality period were unaffected.
(4) In trials, in which oyster seed were brought from all over Japan to Hiroshima Bay, to find those types resistant to mass mortality, the mortality was lowest in those stunted or "hardened" seed which were normally produced for export.

Therefore, instead of studying the causes of mass mortality, the Japanese oyster workers developed hardening techniques in an attempt to produce oysters which were resistant to mass mortality. This approach was of far more value to the oyster culturists than a pure research approach would have been. The hardened seed are prevented from producing heavy gonads and excessive spawning, and oysters on rafts may also be moved to nutritionally poor areas during the gonad maturation period and then back into richer areas in late summer for fattening. These methods are intended to postpone and prevent large gonad development and massive spawning which are considered the primary cause of mass mortalities in conjunction with high temperatures and eutrophication (Koganezawa, 1975).

Mass mortalities in Hiroshima Bay in recent years have been related to the outbreaks of fouling organisms closely related to eutrophication. Attempts to improve the viability and quality of oysters continue against the background of deteriorating water quality which may be slightly arrested by tighter pollution regulations for industry.


## E.  *Water Quality*

Water conditions have changed dramatically in the Hiroshima area in the last 30 years, with one of the basic parameters, transparency, a general guide to the increase in effluents over the years. Since the 1950s average transparency has decreased by 2 m in all regions of the Inland Sea with maximum values now of 10 m with suspended solids ranging from 2 to 19 ppm with maximum values of 33 ppm (Murakami, 1972). In Hiroshima Bay transparency is below 5 m for most of the year with values of > 5 m in November–December. Chemical oxygen demand (COD) values are another useful indicator particularly of organic pollution, and these values which now average 2 ppm (sea water) and 20–25 mg/g (mud) in summer, have doubled in the last 25 years. Dissolved $O_2$ content of sea water ranges from 4·3 to 7·5 ml/l in surface layers and 3·0 to 5·5 ml/l near the bottom (normal levels ranged from 6 to 8 ml/l). Total sulphur content of mud had reached levels of 0·1–1·0 mg/g by the 1970s. As regards nutrient levels, the concentrations of inorganic phosphorus ($PO_4$-P) and inorganic nitrogen (sum of $NH_4$-N, $NO_2$-N, $NO_3$-N) in sea water have increased 5–10 times in the last 30 years with average summer values in the Inland Sea for nitrogen of 2–12 μg-at./l and for phosphorus of 0·5 μg-at./l. In Hiroshima Bay, the concentrations of nitrate-nitrogen increased ten times from 1965 to 1970, with average values remaining around 7–10 μg-at./l, in the 1970s with levels ten times higher occurring in some areas. This was due to a polluting nitrogen load of around 33 tonnes a day in the industrial heyday of the 1960s of which 45% was from industrial waste, 35% from domestic sewage and 20% from agricultural run-off. The average background levels for nitrogen in Japanese bays and inlets is around 2 μg-at./l and the threshold level for the onset of eutrophic conditions is considered to be around 7 μg-at./l of inorganic N, and 0·5 μg-at./l of inorganic P. This would indicate that Hiroshima Bay has been verging on a eutrophic state since the 1960s and indeed serious red tides occurred regularly from 1965 with heavy outbreaks of the eutrophic organism *Euglena* sp. in 1969 and 1970. In the 1970s *Gymnodinium* outbreaks of up to 3000 cells/ml occurred in the Hiroshima area. Apart from these sudden outbreaks of eutrophic organisms, the general species composition of the phytoplankton has changed over the years. Formerly diatoms predominated in the bay but now such organisms as *Noctiluca scintillans* (McCartney) Ehrenb. and primitive euglenoid species such as *Hemieutreptia antiqua* (Hada) prevail, indicative of a chronic eutrophic condition. Diatoms are considered essential for the high taste quality of Hiroshima oysters and inevitably the change in phytoplankton food organisms has resulted in deterioration in the flavour of the oysters. Changes in species composition have also occurred in the benthic communities with recognized pollution indicators such as *Capitella* sp. establishing themselves and a few azoic regions appearing near Kure and Iwakani (Kitamori, 1972). Also the sudden outbreaks of *Hydroides* (see p. 32) and unusual "red water" conditions caused by pinnotherid crabs (Arakawa, 1973c) are due to increasing eutrophication.

Another source of nitrogen is the faecal and pseudofaecal material deposited from rafts which has been calculated at around 300 000 tonnes/year in Hiroshima Bay or 20–30 tonnes/raft/year. This material is composed of 0·1 % total nitrogen (dry weight) and leads to deoxygenation and accumulation of $H_2S$ below the rafts (Arakawa, 1973a). Sea bottom ploughing techniques are being developed to turn over sediments and reoxygenate them by using high pressure hoses alongside the ploughs. Domestic sewage, as well as contributing nutrients, leads to a build-up of coliform bacteria in the Bay which have increased in numbers along with the development of Hiroshima where the population now stands at over 2·5 million. In the early 1960s *Escherichia coli* averaged just over 100 MPN (plate count estimate) in Hiroshima Bay water samples. By the middle 1960s the numbers averaged over 200 MPN and by the 1970s numbers from 250 to 2400 MPN appeared (Arakawa, 1973a). *E. coli* standards for oyster meats were set at < 230, and < 50 000 for total bacteria. In 1976 and 1977 inspections at the major oyster market centres of Tokyo, Yokohama, Nagoya, Kobe, Osaka and Kyoto showed 91.8% of oyster consignments from Hiroshima to be below the maximum *E. coli* count and 82·9% to be below the total bacterial maximum level. Inspections at the production sites before the oysters are shipped to market reveal 80–90% below the maximum level. These are improvements over ten years ago when only 60–70% of oyster samples passed the standard. For marketing purposes, growing areas are classified as safe if the coliform levels are $< 70$ colonies/100 ml of sea water and unsanitary if above this level. Oysters in the latter category are given special purification treatment with chlorinated sea water. The oysters in the former category are washed with "pure" sea water which has been defined as follows: in 100 ml samples the coliform count should be $< 1 \cdot 8$; COD should be $< 2$ppm, DO should be $> 60 \%$; oil should be $< 0 \cdot 1$ ppm; pH should be in the range $5 \cdot 8$–$9 \cdot 0$; heavy metals below normal detection levels (Anon, 1978a).

The factors which have controlled eutrophication in the Hiroshima area are probably the chemical complexing ability of sea water and water exchange which could account for 30–40% of the load. Much tighter anti-pollution legislation also came into effect after the dramatic occurrence of mercury poisoning in Minamata Bay, south-west Kyushu, in 1956. The levels of mercury and cadmium in the Seto Inland Sea mud at that time were 1–2 ppm and $0 \cdot 5$–3 ppm respectively (cf. $12 \cdot 0$–$60 \cdot 0$ ppm methyl mercury in Minamata sediments (Hartung and Dinman, 1972)). Extensive hydrographic investigations aimed at improving the water quality in the Inland Sea have been conducted and such was the concern for this rich fishing environment that in 1971 a National Industrial Research Institute was established to study pollution problems in the Inland Sea. A government financed hydraulic model of the Hiroshima area of the Seto Inland Sea covering some 8900 m² was built. This scale model was used to predict the behaviour of industrial effluents in the simulated tidal patterns of the Inland Sea. Hydrographic experts reckon that to restore the water quality of the Inland Sea will require the removal of about half a billion tons of polluted sea mud, since pollutants once in the ecosystem are recycled through bacterial degradation and synthesis (Murakami, 1972).

## VI. Production, Harvesting and Marketing

### A.  *National Production*

World oyster production according to Japanese statistics is currently around 900 000 tonnes with the United States accounting for 35% of this total, followed by Japan with 25% and Korea with 18%. Actual consumption of oysters is around 60% of the world production for the U.S.A. and 25% for Japan. Table II shows the trends in world production from 1971 to 1978, indicating Korea's rapid rise as a major producer. Japan's oyster production has remained virtually static for the last 20 years due to the fact that in the main production area Hiroshima, where 70% of national production is harvested, further development of the traditional oyster grounds is impossible and the already heavily exploited grounds are being encroached upon by

TABLE II. WORLD OYSTER PRODUCTION TRENDS IN TONNES × 10³ SHELL WEIGHT

| Producer | 1971 | 1972 | 1973 | 1974 | 1975 | 1976 | 1977 | 1978 |
|----------|------|------|------|------|------|------|------|------|
| America | 351 | 327 | 312 | 267 | 314 | 335 | 278 | 314 |
| Japan | 195 | 217 | 230 | 211 | 201 | 226 | 213 | 232 |
| Korea | 54 | 72 | 93 | 65 | 153 | 165 | 161 | 158 |
| France | 34 | 68 | 72 | 74 | 94 | 91 | 112 | 95 |
| Mexico | 44 | 42 | 31 | 33 | 32 | 34 | 30 | 35 |
| Other | 47 | 26 | 55 | 50 | 50 | 52 | 85 | 65 |
| Total | 725 | 752 | 793 | 700 | 844 | 903 | 879 | 899 |

From Anon., 1978a, 1980.

land reclamation schemes. Current production is just over 200 000 tonnes which represents some 34 000 tonnes of shucked meats annually of which some 10–12% are exported in fresh frozen and tinned form. Table I (see p. 4) showed the production of oyster meat in Japan alongside the major Prefecture producer, Hiroshima, for the period 1965–1978. The value of this national harvest is also shown in Fig. 17, amounting to over 24 billion yen or £48 million in 1979, with the value of Hiroshima production more than doubling from 1971 to 1979. Table III summarizes the quantity and type of

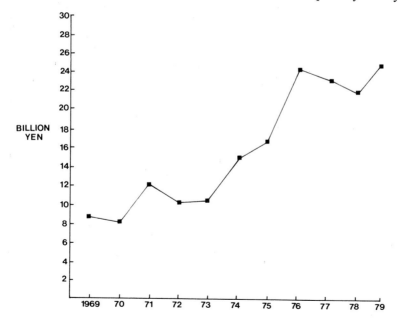

FIG. 17. Japan's national oyster meat production value.

meat exported from Japan in the 1970s and Table IV details the value of fresh and processed meats exported world-wide from Japan in 1979. More than 85% of exported meats are tinned in oil or plain boiled, with fresh frozen meats amounting to 10–12%, and a small percentage are dry preserved meats.

TABLE III. TYPES OF PROCESSED MEAT EXPORTS FROM JAPAN (WEIGHT IN TONNES)

| | Tinned | | | Fresh | Dry | |
|---|---|---|---|---|---|---|
| Year | In oil | Boiled | Other | Frozen | Preserved | Total |
| 1971 | 1801 | 1403 | 38 | 527 | 7 | 3776 |
| 1972 | 1859 | 3443 | 189 | 313 | 104 | 5908 |
| 1973 | 1222 | 2707 | 49 | 458 | 122 | 4558 |
| 1974 | 1272 | 2467 | 35 | 387 | 38 | 4199 |
| 1975 | 1041 | 1257 | 26 | 266 | 29 | 2619 |
| 1976 | 965 | 1490 | 66 | 269 | 20 | 2810 |
| 1977 | 491 | 513 | 83 | 343 | 4 | 1434 |
| 1978 | 922 | 1550 | 37 | 317 | 5 | 2831 |
| 1979 | 756 | 2095 | 23 | 396 | 17 | 3287 |

From Anon., 1980.

The fall of processed exports recently (see Table III) has been due to increased competition from cheaper Korean tinned oyster products which are considered to be inferior in taste and quality to the Japanese processed pro-

TABLE IV. JAPAN'S EXPORT MARKETS FOR PROCESSED MEATS IN 1979

| | Tinned | | | Fresh frozen | Dry smoked | Total | Total value (million |
|---|---|---|---|---|---|---|---|
| Market | In oil (tonnes) | Boiled (tonnes) | Other (tonnes) | (tonnes) | (tonnes) | (tonnes) | yen) |
| America | 244 | 1806 | 2 | 255 | 3 | 2310 | 1473 |
| Canada | 334 | 135 | 0·6 | — | 1 | 470·6 | 455 |
| Australia | 59 | 2 | 10 | 34 | — | 105 | 98 |
| South Africa | 37 | 34 | — | — | — | 71 | 64 |
| New Zealand | 4 | 9 | 5 | 39 | — | 57 | 46 |
| U.K. | 15 | 24 | — | 42 | — | 81 | 61 |
| Hong Kong | 3 | 1 | 0·5 | 3 | 13 | 20·5 | 36·2 |
| Singapore | 2 | 2 | 4 | 11 | — | 19 | 19 |
| Other | 58 | 82 | 0·9 | 12 | — | 152·9 | 138·8 |
| Total | 756 | 2095 | 23 | 396 | 17 | 3287 | 2391[a] |

[a](£4.8 million).
From Anon., 1980.

duct. Very recently, Korean production has fallen due to the limited productivity of Korean waters. The export of fresh frozen oysters from Japan has remained steady, particularly to the U.S.A. The main export to Europe and the U.K. is of oysters tinned in oil with the bulk of this again going to the U.S.A. The value of these processed oysters for export is around 500–800 yen/kg for boiled meats, and 800–1000 yen/kg for fresh frozen according to the country importing, with the smoked and dried oysters in U.S.A. and Hong Kong worth 2000–3000 yen/kg. These national production statistics are more easily appreciated and put in perspective when the individual regional data is analysed and the production figures are related to Prefectures and cooperative bodies. In order to do this, the production data relating to Hiroshima, the major producer, and other Prefectures such as Miyagi, Okayama, and Iwate will be analysed. Table V shows the main Prefectures involved in oyster cultivation and compares the production trends up to 1978. The Hiroshima area therefore supplies 65–76% of Japan's production and also the major oyster market centres with a similar percentage or more of their sales (see pp. 46–50).

TABLE V. NATIONAL OYSTER MEAT PRODUCTION BY PREFECTURE (WEIGHT IN TONNES)

| Prefecture | 1971 | 1972 | 1973 | 1974 | 1975 | 1976 | 1977 | 1978 |
|---|---|---|---|---|---|---|---|---|
| Hiroshima | 27 119 | 23 479 | 24 594 | 23 712 | 23 181 | 22 500 | 25 892 | 22 598 |
| Miyagi | 3774 | 4451 | 3829 | 4236 | 3139 | 4667 | 3867 | 3822 |
| Okayama | 2012 | 2146 | 1936 | 2007 | 2936 | 3254 | 3503 | 3952 |
| Iwate | 832 | 921 | 899 | 929 | 885 | 1160 | 1163 | 1261 |
| Ishikawa | 531 | 567 | 446 | 520 | 613 | 621 | 710 | 664 |
| Mie | 334 | 245 | 401 | 233 | 313 | 271 | 136 | 107 |
| Kagawa | 242 | 253 | 103 | 330 | 310 | 520 | 248 | 200 |
| Hokkaido | 299 | 306 | 322 | 328 | 303 | 300 | 328 | 184 |
| Other | 833 | 739 | 899 | 816 | 921 | 999 | 1358 | 1170 |
| Total | 35 976 | 33 107 | 33 429 | 33 111 | 32 600 | 34 292 | 37 205 | 33 935 |
| Hiroshima (%) | 75·4 | 70·1 | 73·6 | 71·6 | 71·1 | 65·6 | 69·6 | 66·6 |

From Anon., 1980.

## B.  *Hiroshima Production*

Production of oysters in Hiroshima became depressed in the 1970s after the damaging effects of exposure to *Hydroides* in 1969 and 1970, typhoon damage in 1970, cadmium contamination in 1972 and oil spill damage in 1975. After peak production of 31 188 tonnes in 1968, production fell to 14 358 tonnes in 1970 and many cultivators turned to seaweed "nori" production. There were various national economic factors arising in the 1970s such as the oil crises, Japan's strong yen, international pressure on Japan's exports, all of which resulted in the oyster being costlier to produce, with consequent increases in market prices and a reduction in the amount of oysters being consumed in Japanese households. Recently, however, there has been some revival of the industry, with attempts being made to improve oyster quality and the oyster farm environment by restricting the number of rafts. There is also Government pressure in industry towards improving the water quality within the Inland Sea in general which will result in improved oyster quality and revival of the public's interest in eating more oysters after the cadmium scares and unhygienic coliform bacteria levels of the 1960s and early 1970s. The recent problem of the international 200-mile limits has also resulted in Japan looking more towards its inland fisheries and mariculture production and this, together with the inroads being made by cheaper Korean oysters ("Kankoku kai"), has resulted in a campaign to revive and re-establish the Hiroshima oyster as the number one quality oyster in a region which is proud of its 400-year tradition in oyster cultivation. This will be achieved by setting a target production of around 23 000 tonnes of meat, and better management of monthly production at harvest time tied to market demands, and also by what is called sanitary management, to improve public acceptance of Hiroshima meat.

Figure 18 shows the transition in meat production in Hiroshima from the early 1950s. The peak production years of 1965–1968 coincided with the increase in the number of rafts over that period although the yield/raft

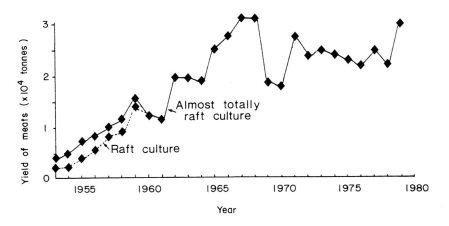

FIG. 18. The transition in Hiroshima oyster meat production (data from Hiroshima Prefecture Experimental Station).

decreased (see Fig. 15). Recently raft numbers have decreased from 11 000 to around 9200–9800 and raft yields have fluctuated between 2·0–2·5 tonnes/ raft, representing around 4 kg of meats/hanging ren (600 ren/raft). The optimum production levels are considered to be around 20 000 tonnes from less than 9000 rafts with ren density reduced to 400 ren/raft. At these levels it is thought water quality will stabilize and the bottom layers recover faster after summer deoxygenation. Table VI details Hiroshima production of fresh and processed meats from 1971 to 1979 with production value, and also shows the number of co-operative bodies and rafts concerned. Raft numbers therefore stabilized below 10 000 in the 1970s and the number of co-operative bodies decreased by 20%. Production had stabilized around 20 000–24 000 tonnes/year but increased again in 1977 with the introduction of more rafts and again in 1979, with the production prediction for 1980 also being high at 25 000–30 000 tonnes. While the co-operatives are being urged to accept a 23 000 tonne/year maximum figure by Hiroshima Prefecture, the Prefecture biologists recommend that a figure of 20 000 tonnes or less would be more ecologically acceptable.

The co-operatives are unsure about setting limits to production in case of natural disasters or a poor growth year, and high production volume planning is a safeguard against such occurrences. The co-operatives' reaction to smaller oysters and lower raft yields is to increase the number of culture facilities, and this short-sighted attitude is likely to continue as oyster values increase. Comparing 1971 and 1979 production, it can be seen that oyster values have doubled in a very short period, especially those for fresh meats (see pp. 46–50 for details of production costs, wholesale prices and retail prices).

The direct effects of deteriorating water quality and over-production on rafts is seen in the yield of meat for individual oysters. The ratio for oyster weight to meat weight was formerly 6 : 1 but now stands at 7 : 1 or 8 : 1. Recently meat yields per oyster have averaged 15–18 g at their peak, as compared to 30 g in the 1950s. Oysters are harvested and shucked from November to April and those of < 15 g are processed in vinegar ("chin mi") at the beginning of the harvest season (November–January). The meats from the largest oysters from February onwards are packed raw for "teppan-yaki" cooking (quick fry). Generally, from October to December demand and prices are high, although taste is not so good. Then from January to March, the oysters fatten with the lower water temperatures ("cold shock"), forming firm flesh. At this time, especially the end of January to the beginning of February the taste is excellent although prices are down with increased supplies. Figure 19 shows meat yield/oyster over the harvesting period for 1975–1979. The rise and fall in meat yield/oyster can be related to the number of rafts operating in those particular years with the highest yields (1975,

TABLE VI. TRENDS IN HIROSHIMA PRODUCTION 1971–1979

|  | 1971 | 1972 | 1973 | 1974 | 1975 | 1976 | 1977 | 1978 | 1979 |
|---|---|---|---|---|---|---|---|---|---|
| Total production (tonnes) | 27 000 | 23 000 | 24 600 | 24 400 | 23 200 | 22 500 | 26 000 | 23 000 | 31 500 |
| Production treatment: | | | | | | | | | |
| Fresh (tonnes) | 14 000 | 10 600 | 16 800 | 15 500 | 15 200 | 15 100 | 14 000 | 15 800 | 17 300 |
| Processed (tonnes) | 13 000 | 12 400 | 7000 | 8900 | 7000 | 7400 | 12 000 | 7200 | 14 200 |
| Percentage processed | 48·1 | 53·9 | 32·5 | 36·5 | 30·2 | 32·9 | 46·2 | 31·3 | 45·0 |
| Total production value[a] | 86·5 | 65·8 | 96·8 | 117·6 | 137·9 | 175·1 | 159·7 | 144·4 | 180·6 |
| No. of operators | 797 | 696 | 689 | 653 | 633 | 624 | 624 | 620 | 630 |
| No of rafts | 9946 | 9723 | 9961 | 9601 | 9289 | 9273 | 9816 | 9367 | 9813 |

[a] × 100 million yen.

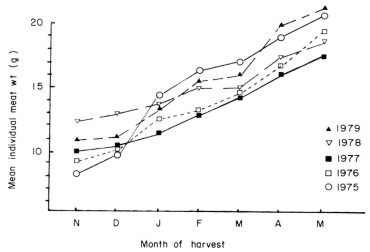

FIG. 19. Mean yields of individual oysters over the main harvest period (November–May) for 1975–1979 (after Anon., 1980).

1976) associated with lower raft numbers (see Table VI). Each co-operative group operates 15–20 rafts, and with the decrease in operators, who now number around 550 (1979), raft numbers will come down to a level which will allow rafts to be moved around more to improve oyster meat quality and allow the Bay bottom to recover from the effects of faecal deposits.

## C.   *Miyagi Production*

Oyster production in Miyagi is second to Hiroshima although it is only 15–20% of Hiroshima production and about 12% of national production (see Table V). Production remains steady averaging 4000 tonnes of meat a year. As mentioned previously, Miyagi does not have the same reputation for producing first class oysters as Hiroshima and many co-operative groups in Miyagi are involved only in spat collection and hardening seed for export, with current average of 2·25 million hardened ren produced each year and 75% exported. These exports are valued at around £3 million. Miyagi differs from Hiroshima in having potential for expansion, since the long line method of oyster culture and spat collection enables installations to be set offshore as well as in the bays. There are about 2000 groups involved in oyster culture in Miyagi with some 800 households involved in hardened seed production. This means there are twice as many oyster culturists in Miyagi as in Hiroshima and when culture installations are considered there are around 3000 rafts in Miyagi (30% of Hiroshima) and some 9000 long lines. Mean production levels in Miyagi are, however, much lower than Hiroshima for the reasons mentioned above (spat collection) and also because oyster culture in Miyagi is part of a multi-economy with long lines and rafts also being used to accommodate scallops and seaweeds. The situation in Hiroshima where the economy of the culturist is based solely on oysters is unique and has remained economically feasible because of the reputation of the products, its price structure and the proximity of two major oyster markets which are the cities of Kobe and Osaka. Table VII compares the changing situations in Miyagi and Hiroshima during the 1970s with regard to production, operators, installations and yields. Hiroshima has seen slight reductions in total production, number of rafts and number of operators, with some improvements in raft yield. Production in Miyagi has increased slightly with a one-third reduction in rafts and an 80% increase in long lines, although the number of operators has fallen by 40%. The total number of rafts and long lines being operated all over Japan have been estimated at 18 432 and 14 244 respectively (Honma, 1980).

A typical example of Miyagi co-operative production would be the Ishinomaki co-operative in Sendai Bay with 260 culturists. Here production is 250 tonnes of meat/year from 130 growers with wholesale value 25 000 yen/ 20 kg packs (£2.50/kg). Miyagi oysters yield meats of 15–20 g and the individual grower will produce about 2 tonnes/year (value 2·5 million yen). The Ishinomaki co-operative also has 130 seed producers who in 1978 produced 800 000 ren of seed (35% of Miyagi production). These ren with spat cost 600 yen/ren and therefore have a value to the producers of 480 million yen (£960 000) or around 3·5 million yen/producer. At the other end of the scale is the small co-operative group in Ogatsu Bay with 20 operators growing scallops and oysters. Annual oyster production is 40 tonnes of meat from 70 long lines of value 50 million yen (1250 yen/kg). Each operator would work 6–7 long lines (3–4 harvested/year) holding 700 000 yen worth of oysters, and therefore have a yearly harvest value of around 2·5 million yen (£5000) from oysters, similar to the larger co-operatives (see also pp. 30–31).

## D.   Harvesting and Marketing

There are seven major oyster market centres in the south of Japan. In order of tonnage handled (1975–1979 average figures) they are: Tokyo (5650), Nagoya (2650), Osaka (2560), Yokohama (1490), Kitakyushu (990), Kobe (700), Kyoto (670), with another three smaller markets in Himeji, Wakayama and Amagasaki (all near Osaka) which handle another 700 tonnes. In any year in particular markets 70–100% of this produce will come from the

TABLE VII. RECENT TRENDS IN OYSTER CULTURE IN HIROSHIMA AND MIYAGI

|  | Hiroshima | Miyagi |
|---|---|---|
| 1970's production | 27 000–23 000 tonnes | 3774–3822 tonnes |
| Percentage of national production | 71% | 11% |
| No of rafts | 9900–9300 | 3000–2000 |
| No of long lines | — | 5000–9000 |
| Yield per raft | 1·5–2·5 tonnes | 0·65–0·5 tonnes |
| Yield per long line | — | 0·95 ton |
| Yield per unit area (m²): Raft | 12·5 kg | 14·2 kg |
| Long line | — | 15·8 kg |
| No. of operators | 795–620 | 3350–2000 |

Data from 1970–1978.

Hiroshima growers with the Tokyo market handling 38% of the total market, Osaka area markets 25% and Yokohama 10%. Nagoya market, however, being nearer the Mie production area handles only a small percentage of

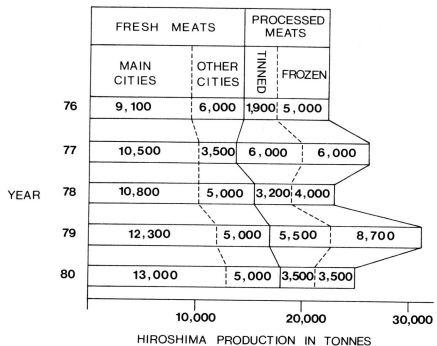

FIG. 20. The production and distribution of fresh and processed oyster meats from Hiroshima (Anon., 1980).

Hiroshima production. Figure 20 shows how each year's production from Hiroshima is classified and distributed (1976–1980). Most of the production is of fresh oyster meat (65 %) destined for the major city markets of Tokyo, Yokohama, Osaka, Kobe and Kyoto. About 35 % of the meat is processed with usually 20 % fresh frozen and about 15 % tinned. Production of fresh meat remains steady with larger volumes of tinned and frozen meat produced when production rises above 25 000 tonnes as in 1977 and 1979. The future trend will be towards marketing of more fresh oyster meat as quality and hygiene are improved.

The growers must pay attention to the timing of the harvesting and the fattening of their stock to catch the top market prices offered in October–December for fresh oysters. Figure 21 shows the shipment of fresh meat from Hiroshima to the major city markets from October to April (1979).

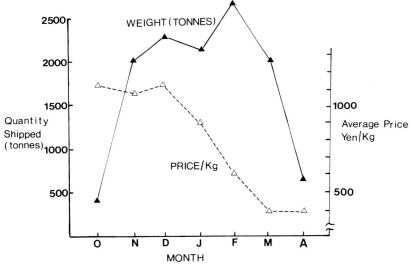

FIG. 21. Monthly shipments of fresh meats from Hiroshima to major city markets in 1979 (tonnage and price/kg) (after Anon., 1980).

Inevitably the larger oysters are available later from March (see Fig. 19) but the bulk of the harvesting (75 %) is from November to February when market prices are higher with 65 % of the harvest sold during November–February. At average prices of 848 yen/kg for fresh and 235 yen/kg for processed oysters (which are low prices) the value of the 1979 harvest to the Hiroshima co-operatives was a record 18 billion yen (£36 million). An analysis of market price movements during the harvest season is seen in Table VIII which compares the major market wholesale values of fresh oyster meats shipped from Hiroshima in 1977. Different markets peak at different times and there

TABLE VIII. MARKET VALUES OF MEATS (YEN/KG) IN THE MAJOR OYSTER MARKETS FOR EACH MONTH OF THE HARVEST SEASON IN 1977

|  | Oct. | Nov. | Dec. | Jan. | Feb. | March | April | Average |
|---|---|---|---|---|---|---|---|---|
| Tokyo | 739 | 1140 | 1265 | 1009 | 765 | 555 | 511 | 887 |
| Yokohama | 732 | 1183 | 1254 | 960 | 705 | 544 | 512 | 877 |
| Kyoto | 1184 | 1172 | 1212 | 960 | 845 | 747 | 664 | 995 |
| Osaka | 1007 | 1083 | 1194 | 879 | 682 | 524 | 504 | 872 |
| Kobe | 1266 | 1000 | 1080 | 912 | 714 | 570 | 410 | 889 |
| Nagoya | 1005 | 744 | 1064 | 967 | 577 | 389 | 405 | 697 |
| Himeji | 1423 | 1167 | 1135 | 1030 | 850 | 1014 | 658 | 1051 |
| Wakayama | 1252 | 1005 | 1185 | 1124 | 685 | 575 | 497 | 977 |
| Kitakyushu | 1147 | 984 | 1089 | 715 | 620 | 539 | 924 | 819 |

From Anon., 1978a.

is a range in prices offered during the peak periods. For example Tokyo and Yokohama markets peaked in December around 1260 yen/kg while Kobe, and Osaka offered 1080 yen and 1194 yen respectively. In general the best prices for the co-operatives are to be had in October–January in Tokyo, Yokohama, Osaka, Himeji and Wakayama. After the peak months, the prices even out over all the markets as they fall off.

It is important to distinguish between the different values of the oyster meats as they are harvested, shipped to market, wholesale valued and then retailed. Table IX attempts to relate first of all production value (estimate by the grower), then market value of the meats (bidding prices), followed by retail value (to the market wholesaler or middle man). These figures are based on average seasonal costs and prices from 1972 to 1979, and the market value can range considerably during the harvest season (see Table VIII). Table IX shows market values rising sharply from 1973 with profit

TABLE IX. HIROSHIMA MEAT VALUES (YEN/KG) DURING PRODUCTION AND MARKETING

|  | 1971 | 1972 | 1973 | 1974 | 1975 | 1976 | 1977 | 1978 | 1979 |
|---|---|---|---|---|---|---|---|---|---|
| Co-operative evaluation | 315 | 328 | 342 | 504 | 411 | 566 | — | 620[a] |  |
| Market average values |  |  |  |  |  |  |  |  |  |
| Fresh | 421 | 440 | 489 | 650 | 715 | 915 | 875 | 811 | 848 |
| Processed | 212 | 155 | 183 | 189 | 287 | 497 | 310 | 226 | 235 |
| Average | 320 | 286 | 390 | 482 | 577 | 778 | 614 | 628 | 572 |
| Average retail prices in Tokyo (Fresh) | — | 1010 | 1100 | 1350 | 1560 | 1610 | 1910 | 1820 | 1550 |

After Anon., 1980.
[a]Estimate.

margins improving considerably for the growers in the last few years. Retail prices also change considerably according to the city and the month. For example in 1979 the peak prices in November were 3000 yen/kg in Tokyo, 2950 yen/kg in Kobe and 3200 yen/kg in Himeji while at the end of the season in April 1980 the retail prices/kg were around 1000 yen in all areas. The average retail prices had risen almost 90 % from 1972 but fell in 1979 due to the large harvest and low prices for processed meats. In the 1980–1981 season, fresh oysters were selling for 2500 yen in Hiroshima in January, although the meats weighed < 15 g. Cold weather in this New Year period however, means a good market for fresh meats since the public want to eat "kaki nabe" (oyster hot pot) in response to the cold weather. Despite rising oyster prices of 50–90 % in the last ten years, the consumption of fresh oysters has risen steadily to an average 1070 g of meat/household/year and this consumption is over a short 3–4 month winter period.

After shipment of fresh oyster meats from November to March, the meats for processing are marketed from February to June. Processing for the home market involves smoked oysters in tins, boiled oysters in tins, frozen oyster meats and the popular "chin mi" or vinegar-preserved oysters for bar snacks. In 1979, which was a good production year in Hiroshima of 31 500 tonnes, some 14 000 tonnes were processed (see Fig. 20). From this processing, 421 000 tins of smoked meats and 432 000 tins of boiled meats were produced and over 2000 tonnes of boiled meats were exported. Another 6820 tonnes were frozen and 7 tonnes of "chin mi" were produced. "Chin mi" products amounted to 118 tonnes in 1977, but recently dried oyster processing has taken over and has risen from 330 tonnes in 1978 to over 1000 tonnes in 1979. In Hiroshima area there are now 36 small companies involved in the processing of meats of which 12 are freezer businesses, eight are canners, and four produce "chin mi" products, and 17 prepare dried oysters.

Finally one aspect of marketing which has received much attention in recent years concerns the quality and hygiene standards of raw oyster meats before shipment and while in the market. In the production areas, purification by natural sea water washing and by chlorinated water is advised. Growing areas are classified as safe or unsanitary according to whether coliform bacteria are present at levels less than 70 colonies/100 ml or over, respectively. Oysters from safe areas have to be washed with clean (safe) sea water and chlorinated sea water and allowed to purify themselves in inflowing clean sea water. In the 12 m³ purifying tanks, live oysters are held at a density of 25 kg/m³ (300 shells/m³) for 10 h. Oysters from unhygienic areas may spend up to 18–24 h in 3 ppm chlorinated sea water. U.V. irradiated water is sometimes used as well as ozone which is a more efficient sterilizer. After shucking, the oysters are washed in "pure" sea water (for definition see p. 38) and it is advised that the shells are disposed of immediately

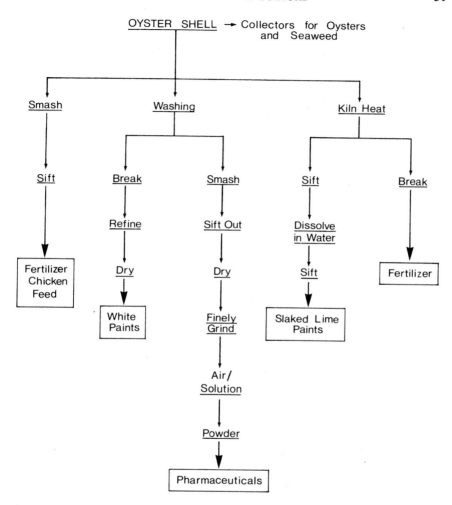

FIG. 22. The utilization of oyster shells after shucking (after Arakawa *et al.*, 1977b).

and not stacked in the same building as the shucked meats. Oyster shells are utilized in many different ways, directly as spat collectors or collectors for seaweeds, and processed as shown in Fig. 22 for use in paint, fertilizer and pharmaceutical products (Arakawa *et al.*, 1977b). All equipment and working surfaces must be kept washed with clean sea water and all employees working with oyster meats must have a medical and faecal examination. Any old equipment such as shells, rafts and racks must be disposed of and not allowed to lie around the shucking beds. After washing and packing, the shucked meats must be stored in ambient temperatures $< 5°C$, so that the meat temperature does not rise above $10°C$ during storage and transport.

The Japanese consumer therefore receives a high quality product, subjected to the highest hygiene standards possible, which is consumed safely in season in most bars, restaurants and many households in the major cities.

## VII. The Future

World oyster production has increased by 25% in the last ten years approaching 0·9 million tonnes (see Table II). World oyster demand is increasing also and is expected to reach over 2 million tonnes in the next two decades (Glude, 1976). Although Japan's production is unlikely to increase substantially in that period, economic growth of the industry is expected to take place due to the home and export market demands. Some expansion of culture grounds would be possible along the Sanriku coasts of Miyagi and Iwate using coastal long line installations. Demand for Japanese spat is likely to continue even though transport costs of cases of spat from Japan have risen steeply. Also hatcheries in America remain expensive to operate. The large American West Coast hatcheries produce a case of spat on oyster shell cultch for between $40–50 while a Japanese case from Miyagi, containing 1000 pieces of cultch and around 10 000 spat has a CIF price of $42–43. The French industry, hit by disease problems in native stock is likely to increase its Pacific oyster output partly with imported Japanese seed.

To the east of Hiroshima is the Akitsu area which is thought suitable for the expansion of culture of the Japanese flat oyster *Crassostrea denselamellosa* (Lischke) and *Crassostrea echinata* (Quoyand and Gaimard). Akitsu in the Aki-Nada sea has higher oceanic salinity and lower nutrient levels than Hiroshima Bay, which render it unsuitable for *C. gigas*. These species may eventually be produced from the first commercial scale Japanese oyster hatcheries together with other marketable *Crassostrea* species. The purpose of these hatcheries will be to provide seed by early summer for ongrowing to commercial size by December (no hardening) and thus re-establish the one-year culture system in Hiroshima Bay. These oysters would then have high market value (Arakawa, personal communication).

The biggest problem facing the further development of the Japanese oyster industry in the traditional areas has been termed "misshoku", close culture or over-culture. This tendency of the growers to crowd culture areas and to overload rafts in response to diminishing meat yields is being heavily criticized by the Prefecture laboratories. Present government policy is also in favour of expanding Hiroshima's industry and population, and the subsequent land reclamation and effluent loads will be detrimental to the oyster industry.

The Japanese methods of oyster farming are highly productive and cost effective principally due to the skilled labour availability and home market demand. Income for oyster growers has risen steadily and tripled in the last ten years because of the rise in oyster market values. Annual family incomes for oyster growers in the Hiroshima area were estimated at around 15 000 000 yen (£35 000) in the late 'seventies, making them the most prosperous of the rural/coastal socio-economic groups (Anon., 1978b). With such a labour intensive system, any shortage of labour is a serious problem, such as lack of recruitment of young workers. The number of younger workers (18–40 year old group) involved in oyster culture is 2–3 times that found in other fishing work because of the heavy nature of the work and the amount of labour involved. Also certain operations such as collector preparation and shucking may involve more than 35% female labour. An oyster group will spend a total 14 000 man hours/year in production as compared to 2350 man hours for inshore trawling and 3200 man hours for "nori" cultivation (Anon., 1978b). Many oyster culturists beset by culture problems (in the early 'seventies) returned to "nori" cultivation. "Nori" cultivation is more automated and the profits are considerable in good years although yearly fluctuations in harvest and quality are characteristic of this culture (1981 saw a boom in exports to the U.S.A. of 586 million yen).

The Japanese hanging culture techniques have been developed as efficiently as possible with high production in mind. On an hourly work rate basis, oyster culture returns between 1100–1200 yen/h which is twice the return for "nori" culture, while inshore trawling returns 1000 yen/h but is seasonal (Anon., 1978b). Japanese oyster production and yield are said to be the highest in the world, approaching 58 tonnes of meat/ha/year (Fujiya, 1970). The above data are based on individual raft yields in specific areas however, whereas it would be more realistic to calculate yield estimates based on the total sea area utilized for culture. Recent regional data would suggest much lower productivity. For example in 1979, Hiroshima produced over 85% of Japan's production at a record 31 500 tonnes of meat (see Table VI). This production came from 3500 ha of culture ground (including rack areas) which would indicate an average yield of 9 tonnes of meat/ha/year. Likewise Miyagi's recent average production of 3800 tonnes of meat is produced from 2300 ha with most of this area being used for hardening on racks. Average regional yield is probably in the region of 4 tonnes of meat/ha/year. The raft and long line yields in Table VII, based on unit yield per culture facility, indicate Miyagi's installations as being more productive than those of Hiroshima, although available culture area is obviously being more heavily exploited in Hiroshima Bay. In other parts of the world, where hanging culture is not normally practised, production is still impressive. In the Charente Maritime region of France, production of over 25 000 tonnes/

year (in shell) of the Pacific oyster was possible from 2300 ha of oyster parks (Korringa, 1976). Best French yield has been calculated at 0·94 tonnes/ha/ year and in the U.S.A. at 5 tonnes/ha/year (Fujiya, 1970).

The sophistication of the Japanese system of production has been questioned and criticized as not producing shells of good shape and size as required in other countries where the half shell trade predominates (Korringa, 1976). However, the Japanese possess an acute sense of taste, and their production of meats of a particular taste quality by careful management of fattening and harvesting in selected areas is highly organized. They also demonstrate "sophistication" in their promotion and marketing expertise and their attention to high standards of hygiene. It is this author's opinion that the Japanese's detailed knowledge of their species and culture areas and the interaction between the two is more developed than in other countries.

## VIII. Acknowledgements

This review is based on information gathered on trips to Hiroshima and Miyagi in 1978, 1979 and 1980, as a recipient of a Monbusho grant from the Japanese Ministry of Education. I am grateful to Professor Uno of the Tokyo University of Fisheries for general advice on the industry. I am indebted to Dr. K. Arakawa of the Hiroshima Prefecture Experimental Station, Ondo Island with whom all aspects of the industry were discussed. Thanks are also due to his staff and members of the Hiroshima Co-operative for detailed discussions on commercial aspects of the system.

In Miyagi my thanks are due to Mr. Sasaki and Dr. Thukuda of Ishinomaki Experimental Station for information and tours of the Sendai Bay area and also to the Ishinomaki Co-operative for commercial discussions.

## IX. References

Anon. (1978a). Hiroshima oyster production and marketing in 1978. *Hiroshima Prefecture Leaflet.* (In Japanese.)

Anon. (1978b). "Movements of Fishery Industries in Hiroshima (1976 to 1978)." Nohrin Tohkei Kyokai, Hiroshima. (In Japanese.)

Anon. (1980). Hiroshima oyster production and marketing in 1980. *Hiroshima Prefecture Leaflet.* (In Japanese.)

Arakawa, K. Y. (1971). Notes on a serious damage to cultured oyster crops in Hiroshima caused by a unique and unprecedented outbreak of a serpulid worm, *Hydroides norvegica* (Gunnerus) in 1969. *Venus—Japanese Journal of Malacology* Vol. 30 **No. 2,** 75–83. (In Japanese.)

Arakawa, K. Y. (1973a). Aspects of eutrophication in Hiroshima Bay viewed from transition of cultured oyster production and succession of marine biotic communities. *Bulletin of the Hiroshima Fisheries Experimental Station* **No. 2,** 43–48. (In Japanese.)

Arakawa, K. Y. (1973b). "Prevention and Removal of Fouling on Cultured Oysters. A Handbook for Growers." (Translated into English by R. B. Gillmor, Department of Oceanography, University of Maine. *Maine Sea Grant Technical Report* **56**, 1–38, 1980.)

Arakawa, K. Y. (1973c). Notes on an unusual occurrence of red water caused by a shoal of pinnotherid crab, *Tritodynamia horvanthi* (Nobile), off the coast of Hiroshima Prefecture in 1971. *Bulletin of the Hiroshima Fisheries Experimental Station* No. **4**, 129–131. (In Japanese.)

Arakawa, K. Y. (1974). Notes on damage to cultured oyster crop in the vicinity of Ondo, Hiroshima, caused by keen competition with blue mussel, *Mytilus edulis galloprovincialis* (Lamarck). *Bulletin of the Hiroshima Fisheries Experimental Station* No. **5**, 35–37. (In Japanese.)

Arakawa, K. Y. and Kubota, H. (1973). Biological studies on prevention and extermination of fouling organisms attached to cultured oysters—I. Distribution and seasonal settlement of *Hydroides norvegica* (Gunnerus) in Hiroshima Bay. *Bulletin of the Hiroshima Fisheries Experimental Station* No. **4**, 13–17. (In Japanese.)

Arakawa, K. Y., Hoshina, T. and Matsuzato, T. (1976). An annotated bibliography of diseases in oysters. *Bulletin of the Hiroshima Fisheries Experimental Station* Nos **6–7**, 75–101. (In Japanese.)

Arakawa, K. Y., Hoshina, T. and Matsuzato, T. (1977a). A bibliography of enemies of oysters with annotations. *Bulletin of the Hiroshima Fisheries Experimental Station* No. **8**, 27–52. (In Japanese.)

Arakawa, K. Y., Tsuchiya, N., Kaneyasu, T. and Tanimoto, T. (1977b). Annotated bibliography of utilisation of oyster shells (by-product of the oyster culture industry). *Bulletin of the Hiroshima Fisheries Experimental Station* No. **8**, 56–63. (In Japanese.)

Cahn, A. R. (1950). Oyster culture in Japan. (Researched by T. Ino.) *General Headquarters Supreme Commander for the Allied Forces Natural Resources Section Report* No. **134**, 80 pp., Tokyo, 1950. (Also in: *U.S. Fisheries and Wildlife Service Leaflet* **383**, 80 pp.)

Fujiya, M. (1970). Oyster farming in Japan. *Helgoländer Wissenschaftliche Meeresuntersuchungen* **20**, 464–479.

Gillmor, R. B. and Arakawa, K. Y. (1976). Some field experiments on setting in *Crassostrea gigas* (Thunberg). *Bulletin of the Hiroshima Fisheries Experimental Station* Nos **6–7**, 37–51. (English translation.)

Glude, J. B. (1976). "Oyster Culture. A World Review." FAO Technical Conference on Aquaculture, Kyoto, Japan, 1976. FIR: AQ/Conf/76/R.16. FAO, Rome.

Hartung, R. and Dinman, B. D. (1972). "Environmental Mercury Contamination". A. Arbor, Michigan.

Honma, A. (1980). "Aquaculture in Japan." Japan FAO Association, Tokyo. (In English.)

Imai, T. and Hatanaka, M. (1949). On the artificial propagation of the Japanese common oyster, *Ostrea gigas* Thunberg by naked flagellates. *Bulletin of the Institute of Agricultural Resources, Tohoku University*, Vol. 1 No. **1**, 1–7. (In Japanese.)

Imai, T. and Sakai, S. (1961). Study on breeding the Japanese oyster, *Crassostrea gigas*. *Tohoku Journal of Agricultural Research* **12**(2), 125–171. (In Japanese.)

Imai, T., Numachi, K., Gizumi, S. and Sato, S. (1965). Research on the high mortality of oysters in Matsushima Bay—II. Research on the cause of death by transplantation test and the study on preventative measures. *Research Report of Tohoku Fishery Research Institute* No. **25**, 27–38. (In Japanese.)

Kanno, H., Sasaki, M., Sakurai, Y., Watanabe, T. and Suzuki, K. (1965). Research on the high mortality of oysters in Matsushima Bay—I. Situation of high mortality and environment. *Research Report of Tohoku Fishery Research Institute* No. 25, 1–26. (In Japanese.)

Kitamori, R. (1972). Faunal and floral changes by pollution in the coastal waters of Japan. *Proceedings of the International Ocean Development Conference. Tokyo.* 1972, 71–77. (In English.)

Koganezawa, A. (1972). Ecological studies at a seed oyster production area—II. Studies on the spawning condition of the Pacific oyster in Matsushima Bay. *Bulletin of the Japanese Society of Scientific Fisheries* Vol. 38(12), 1315–1324. (In Japanese.)

Koganezawa, A. (1975). Present status of studies on the mass mortality of cultured oysters in Japan and its prevention. *Proceedings of the Third U.S.-Japan Meeting on Aquaculture, Tokyo*, 1974, 29–34. (In. English.)

Koganezawa, A. (1976). The status of Pacific oyster culture in Japan. *FAO Technical Conference on Aquaculture, Kyoto, Japan* 1976, FIR: AQ/Conf/76/E.69. (In English.)

Koganezawa, A. (1978). Ecological study of the production of seeds of the Pacific oyster, *Crassostrea gigas*. *Bulletin of the Japan Sea Regional Fisheries Research Laboratory* No. 29, 1–88. (English summary.)

Koganezawa, A. and Goto, K. (1972). Ecological studies at a seed oyster production area—I. Characteristics of mother oyster populations in Sendai Bay. *Bulletin of the Japanese Society of Scientific Fisheries*, Vol. 38 No. 1, 1–8. (In Japanese.)

Koganezawa, A. and Ishida, N. (1973). Ecological studies at a seed oyster production area—III. Distribution of oyster larvae in the northern area of Sendai Bay. *Bulletin of the Japanese Society of Scientific Fisheries*, Vol. 39 No. 2, 131–147. (In Japanese.)

Korringa, P. (1976). Farming the cupped oysters of the genus *Crassotrea*. In "Developments in Aquaculture and Fisheries Science", Vol. 2, Elsevier, Amsterdam.

Matsuzato, T., Hoshina, T., Arakawa, K. Y. and Masumura, K. (1977). Studies on the so-called abnormal egg-mass of Japanese oyster, *Crassostrea gigas* (Thunberg) —I. Distribution of the oyster collected in the coast of Hiroshima Pref. and parasite in the egg-cell. *Bulletin of the Hiroshima Fisheries Experimental Station* No. 8, 9–25. (In Japanese.)

Murakami, A. (1972). Marine pollution in the Seto Inland Sea. *Proceedings of the 2nd International Ocean Development Conference, Tokyo*, 1972, 183–188. (In English.)

Ogasawara, Y. (1972a). Basic studies on method of farming oysters—Part 1. Distribution of larvae. *La Mer. Bulletin de la Societé Franco-japonaise d'oceanographie*, Vol. 10 No. 2, 56–69. (English translation.)

Ogasawara, Y. (1972b). Basic studies on method of farming oysters—Part 2. Spat collection. *Proceedings of the 2nd International Ocean Development Conference, Tokyo*, 1972, 1665–1691. (English translation.)

Ogasawara, Y., Kobayashi, U., Okamoto, R., Furukawa, A., Kisaoka, M. and Nogami, K. (1962). The use of the hardened seed oyster in the culture of the food oyster and its significance to the oyster culture industry. *Bulletin of Naikai Regional Fisheries Research Laboratory* No. 19, 1–13. (English translation.)

Seno, H. (1938). Review of the development of oyster culture in Japan. *Science* (Kagaku) Vol. No. 6, 240–250. (In Japanese.)

Seno, H. and Hori, J. (1927). A new method of fattening oysters. *Journal of the Imperial Fisheries Institute*, Vol. 22 **No. 4,** 69–72. (In Japanese.)

Watanabe, T., Nishibori, S., Abe, K. and Ohta, H. (1972). The present situation of water pollution, and the relationship between the pollution of bottom material and movement of sea water as of the latter part of the 1960's of Matsushima Bay. *Research Report of the Fisheries Experimental Station of Miyagi Prefecture*, Vol. 7, Part VII, 53–63. (English translation.)

Wisely, B., Okamoto, R. and Reid, B. L. (1978). Pacific oyster (*Crassostrea gigas*) spatfall prediction at Hiroshima, Japan, 1977. *Aquaculture*, Vol. 15 **No. 3,** 227–241.

# Marine Toxins and Venomous and Poisonous Marine Plants and Animals (Invertebrates)

## F. E. Russell

*College of Pharmacy, University of Arizona,
Tucson, Arizona, U.S.A.*

ADVANCES IN MARINE BIOLOGY VOL 21
ISBN 0–12–026121–9

# I. Introduction

The venomous and poisonous marine animals have held a fascination for man since the very beginnings of written history. Early peoples often attributed the consequences of their bites and stings, or their ingestion, to forces beyond nature, sometimes to vengeful deities thought to be embodied in the animals themselves. To these early peoples, the effects of bites or stings by animals were so surprising, so varied, and often so violent that the injuries were usually shrouded with much myth and superstition. Even today, considerable folklore about these animals still exists. The task of separating fact from fiction is often a formidable one, and one not always lightened by the passage of time.

The study of the marine toxins, *marine toxinology*, has also enjoyed a fascinating history, since by the very nature of the complexity of these substances and their ability to destroy life by complicated and sometimes undetermined means, their investigation has invited speculation, exaggeration and sometimes pure fantasy. However, during the past three decades considerable progress has been made on the chemistry and physiopharmacology of the marine toxins, as reflected in the fine review works of Phillips and Brady (1953), Kaiser and Michl (1958), Halstead (1965–1970), Baslow (1969), Martin and Padilla (1973), Scheuer (1973–1981), Southcott (1979) and Hashimoto (1979), as well as our own lesser works (Russell, 1965a, 1971a, b). Recently, Halstead (1978) has also abridged his three-volume compendium and added some new data. Of particular importance to marine toxinology have been the various proceedings of the International Society on Toxinology (1967–1980), as well as the numerous excellent publications on marine animals appearing in the journal *Toxicon*.

Previously, I suggested that approximately 1000 species of marine organisms are known to be venomous or poisonous (Russell, 1965a). From the reviews noted above, particularly that of Hashimoto (1979) and the volumes of Scheuer (1973–1981), as well as data extracted from a bibliography on venomous and poisonous marine animals and their toxins (Russell *et al.*, in prep.), it would be fair to say that the number of such marine organisms must now be in excess of 1200. For the most part, these animals are widely distributed throughout the marine fauna from the unicellular protistan, *Gonyaulax*, to certain of the chordates. They are found in almost all seas and oceans of the world. In most areas they do not constitute a medical or socioeconomic problem but in a few scattered regions, such as the South Pacific, where ciguatera poisoning sometimes gives rise to serious public health and economic problems, and in the case of paralytic shellfish poisoning, the poisonous marine animals have presented a threat to man's health and economy.

The present review treats the toxins of some of the more venomous and poisonous organisms of the world. For the most part, it is concerned with the chemical, toxicological and immunological properties of the toxins, the animal's venom apparatus and the mechanism of envenomation. Some attention has been given to the general biology of the animals and to the clinical problem of poisoning in man. A second purpose of the review is to present an account of several of the more interesting current problems in marine toxinology.

## A. *History and Folklore*

According to Halstead (1965), one of the first descriptions of a poisonous marine animal is found in the hieroglyphics of the tombs of Ti (Fifth Dynasty, *c.* 2700 B.C.), where the poisonous pufferfish, *Tetraodon stellatus*, is seen represented by the hieroglyphic inscription *crept*, *špt* or *shepet*. Its toxic properties appear to have been known even at that time. In Exodus 7:20–21 (*c.* 1500 B.C.), Moses records a toxic dinoflagellate bloom: "... and all the waters that were in the river turned to blood. And the fish that was in the river died; and the river stank, and the Egyptians could not drink of the water of the river".

The Greek physician-poet Nicander (384–322 B.C.) notes the venomousness of the stingray and likens the moray eel to the viper because "its bite poisons with a venom like that of the viper". Amusingly, he wrote that some believe that eels may leave the sea, driven by desire, to copulate with vipers.

However, the most exhaustive and credulous early writer of natural and medical history, fact and fiction, was Gaius Plinius Secundus (A.D. 23–79), whose voluminous *Historia Naturalis* contains numerous fascinating accounts of venomous and poisonous animals, including sections on the weeverfish and stingray. Describing the stingray, he writes:

> So venimous it is, that if it be struchen into the root of a tree, it killeth it: it is able to pierce a good cuirace or jacke of buffe, or such like, as if it were an arrow shot or a dart launched: but besides the force and power that it hath that way answerable to iron and steele, the wound that it maketh, it is therewith poisoned.

In his *Theriaca* Nicander notes the weeverfish, *Trachinus*, as the sea dragon. So frequently quoted were Pliny's works that Hulme (1895) commented:

> Several writers of antiquity influenced the mediaeval authors, but it is scarcely necessary to detail their labours at any length, since if they lived before Pliny he borrowed from them, and if they lived afterward they borrowed from him, so that we practically in Pliny get the pith and cream of all.

As Klauber (1956) has so aptly concluded, "Pliny's *Historia Naturalis* was the funnel through which we can watch the ancient folklore pouring down into the mediaeval and modern worlds".

In Read's (1939) *Chinese Materia Medica of Fish Drugs*, based on the writings of early Chinese workers (A.D. 618–1644), one can find a list of many venomous and poisonous fishes known to be dangerous to man. Halstead believes that the mention of the yellowtail or amberjack, *Seriola*, during the T'ang Dynasty (A.D. 618–907) as "a large poisonous fish fatally toxic to man" is the first description of a ciguateric fish. Pufferfish poisoning was known as early as the Sung Dynasty (A.D. 960). The caudal spine of the stingray was used in the Orient as a spear head during the times of the Ainu, 2000 years ago, and perhaps before that by the Jomon people (5000–2000 B.C.), who may have also used stingray poison as a powder or paste on their stingray spear heads (Tomlin, 1966; Bisset, 1976). Tokunai (1790) noted that the Ainu used the dried caudal spine of the Japanese stingray, *Dasyatis akajei* (Müller and Henle), with the integumentary sheath intact, as a weapon.

The spines also had a more exotic use in sorcery. According to Natori (1972), if you wished to attract a particular charming girl, you put stingray spines in her footprints as she walked through the snow. This caused her to suffer a severe headache. You then told her what you had done, whereupon she turned and appealed to you to rid her of her headache.

Most writings on marine animals following the Greek–Roman period tended to follow the descriptions of Nicander, Aristotle, Galenus, and Pliny. During the mediaeval ages, Avicenna's (A.D. 980–1036) *Q'anum* contained a good description of venomous and poisonous animal bites and stings and their treatment, and Peter of Abanos' (*c*. A.D. 1250–1316) *De Venenis* contains some interesting entries on marine animal injuries. Although Gazio (1450–1530), Peter Martyr (1457–1526), Rondelet (1507–1557), and Gesner (1516–1565) described various dangerous marine animals in their writings, the most important work of the Renaissance was Jacques Grevin's *Deux livres des Venins* (1568). Grevin characterized the stingrays, weeverfish and sea hares, among others, and attempted to put the various animals into some categorical system. Grevin's work was followed by the contributions of Autenreith (1833), Bottard (1889), Phisalix (1922), and Pawlowsky (1927), among others.

The modern period must begin with the text of Kaiser and Michl (1958), who accumulated the literature to that date and attempted to put order into a nomenclature for animal venoms. Since then, the number of published texts, reviews and articles on the marine toxins has dramatically increased. In preparing our marine bibliography (Russell *et al.*, in prep.), we found that more papers had been written on marine toxins since 1960 than in the previous course of history. In 1968, in the foreword for Halstead's second volume of *Poisonous and Venomous Marine Animals of the World*, I wrote:

The time is nearly past when one scholar can collect and record the writings of a single area of science. Within the next decade the literature on the venomous and poisonous animals will grow far beyond the grasp of any one scholar and perhaps beyond the scope of even a few.

That time has come for the venomous and poisonous marine animals and their toxins, and although this treatise is an attempt at compromise in fullness of treatment, diversity of subjects covered and length, and although I have made every effort to include all important contributions, I am sure the text entertains shortcomings, for which I apologize.

## B. *Definitions*

Every science, and most human pursuits, have words or terms that are peculiar to its confines. *Toxinology*, the study of the naturally-occurring poisons is no exception. In marine toxinology there are quite a number of words commonly employed by workers within the discipline to convey a specific point or item. The term *venomous animals* is usually applied to those creatures which are capable of producing a poison in a highly developed secretory organ or group of cells, and which can deliver this toxin during a biting or stinging act. *Poisonous animals* are generally regarded to be those whose tissues, either in part or in their entirety are toxic. In reality, all venomous animals are poisonous but not all poisonous animals are venomous. Animals in which a definite venom apparatus is present are sometimes called *phanerotoxic* (Gk. φανερόσ, evident + τοξικόν, poison), while animals whose body tissues are toxic are called *cryptotoxic* (Gk. κρυπτόσ hidden). The rattlesnake, stingray and black widow spider are venomous or phanerotoxic animals, while the blister beetles, certain puffer fishes and toads are said to be poisonous or cryptotoxic (Russell, 1965a).

Although the terms *venomous* and *poisonous* are often used synonymously, most investigators have tried to confine the use of the term venomous animals to those creatures having a gland or group of highly specialized secretory cells, a venom duct (although this is not a constant finding), and a structure for delivering the venom. While there has been a tendency to employ the term *venom apparatus* to denote only the sting, spine, jaw, tooth or fang used by the animal to inject or deliver its venom, most biologists now use the term in its broader context, that is, to denote the gland and duct in addition to the sting or fang. Poisonous animals, as distinguished from venomous animals, have no such apparatus; poisoning by these forms usually takes place through ingestion (Russell, 1965a).

The words *toxicity* and *lethality* are frequently interchanged. Historically, the terms were often interchanged. Early workers in both pharmacology and toxicology used the word toxicity whenever they meant lethality, and some textbooks in toxicology still retain the word toxicity in place of lethality. However, most toxinologists try to differentiate between the two. In my opening address before the International Society on Toxinology in 1962, I indicated the more broad usage of the word toxicity:

> A poison that causes a large ulcerating lesion or other tissue defect or a response of a deleterious nature is certainly toxic, while not necessarily lethal, and such pharmacological activities should not be overlooked. Toxicity might also include changes in the blood cells, coagulation defects, changes in nerve–muscle transmission, or any of a number of other deleterious changes that do not necessarily end in death. Certainly, when one uses the word lethality he is being specific; that is, he is defining the end point as death, and only that end point. Such statements as "copperhead venom is not toxic" are unfortunate, for aside from the fact that the venom of *Agkistrodon contortrix* produces tissue changes, it is also lethal, even though the $LD_{50}$ is much higher than for the venom for *Crotalus* species (Russell, 1980b).

In speaking about a venom, therefore, it is important that its toxic and/or lethal properties be carefully defined. Even though some fractions of marine venom may not be lethal, their effects may be of significant physiopharmacological or clinical importance to warrant specific definition.

The $LD_{50}$ (lethal median dose) is defined as that dose of venom, or a venom component, that kills 50% of a given group of animals. Theoretically, the figure is based on data from 100 animals but most investigators generally use 20–30 animals for determination of the $LD_{50}$. They then employ one of several acceptable statistical methods to calculate the $LD_{50}$. The animals are injected with varying doses, ranging from the minimum amount that kills all animals in a group to the maximum amount that kills none. The minimum number of animals in a group is usually consistent and may be as few as four or as many as ten, for reproducible data. The error associated with determination of the $LD_{50}$ is smaller than the error for any other estimated dose of the quantal dose-response curve (Russell, 1966a).

Although there are other definable and undefinable end points for determining the lethal dose of a toxin, and some of these are occasionally employed, they are less satisfactory and more susceptible to error than the $LD_{50}$. Some of the more common but ill-defined end points for lethality are the LD (lethal dose), MLD (minimum lethal dose), $LD_{99}$ (dose that kills 99% of the animals), ALD (approximate lethal dose), $LD_1$, $LD_{100_x2}$, and CLD (complete lethal dose). Although I have quoted these kinds of doses as determined by the various investigators, the reader should be extremely cautious about their significance and even more so about their application in comparative pharmacology or toxinology.

$ED_{50}$ means median effective dose. This is generally employed in two different ways. It is used when the investigator needs a reliable end point for a specific response, such as a transient electrocardiographic change or a muscle twitch, or even death (in which case the $ED_{50}$ and $LD_{50}$ become the same). It is also used in place of the $LD_{50}$. The term *standard dose* is sometimes used to mean the $LD_{50}$, once the measurement has been established. It is frequently employed for pooled lots of a poison from which the original sample was taken. The *standard safety dose* is defined as the percentage increase of a dose above the therapeutic dose that is lethal to a given group of animals or subjects.

When studying a specific response of some marine venoms, the *survival time* of the animal may be measured. In such experiments each animal in the test provides its own numerical estimation of the toxicity of the drug. In order to obtain a graded response on survival time, it is essential that all the animals in the experiment die; this is sometimes overlooked.

Halstead (1978) has proposed the word *ichthyocrinotoxic* for those fishes which produce a poison through glandular secretions not associated with a venom apparatus. This word might be used for the soapfishes, certain gobies, some cyclostomes, boxfishes, toadfishes, lampreys and hagfishes, and some sponges, bryozoans and echinoderms which may release toxic skin secretions into the water, perhaps under stressful conditions. Fish poisoning is synonymous with *ichthyotoxism*, and while it implies any type of poisoning it should probably not be used for those poisonings that involve bacterial contamination.

Halstead (1964) divided the ichthyotoxic fishes into three subdivisions:

(a) *Ichthyosarcotoxic*. Those fishes which contain a toxin within their musculature, viscera or skin, which when ingested produce deleterious effects. This type of poisoning is generally identified with the kind of fish involved: elasmobranch, chimaeroid, clupeoid, ciguatera, tetraodon, scombroid, etc.; it also includes hallucinatory fish poisoning.

(b) *Ichthyootoxic*. Those fishes which produce a toxin that is related to gonadal activity. Most members of this subdivision are freshwater species. This group would include those fishes whose roe is poisonous.

(c) *Ichthyohaemotoxic*. Those fishes which have a toxin in their blood. Some freshwater eels and several marine fishes make up this group (Russell, 1965a).

The word *bloom* is used to indicate an excessive or massive local accumulation of algae or protozoa (protistan), which often causes discolouration of surface water.

Some investigators use the word *ichthyotoxin* or *ichthyotoxic* to denote that a certain substance is poisonous *to* fish. Equally, the word is applied to identify a toxin or toxic substance taken from a fish. In this treatise I have used the word in its latter sense. When a substance kills or is toxic to fish, I have noted this as such.

I regret that I cannot agree with my good friend, Bruce Halstead, concerning the use of certain words or terms used in our science and commonly or uncommonly employed in the literature. My chief difficulty is with the word *biotoxins* and our differences over the use of the word *toxinology*. Firstly, in 1962 I thought I had coined a new word "toxinology", which seemed appropriate for our developing science, the study of poisons derived from the tissues of plants and animals. The word was adopted by the International Society on Toxinology and the journal *Toxicon* in 1963. The word thence found common use in the literature. Some curricula and course work became identified as "toxinology," and subdivisions soon appeared: marine toxinology, reptile toxinology and arthropod toxinology.

Etymology of *toxinology* indicates it is consistent with the principles set forth in *The Oxford English Dictionary* (1933) and compatible with common usage: biology, pharmacology, toxicology, etc. Further, I prefer the word to *biotoxicology* (although this word is certainly not inappropriate), for among other things, it is apparent that toxinology is rapidly becoming an independent science and one which deals specifically with the actions of animal and plant poisons and venoms, and quite separately from the interests of toxicology. Within two decades I visualize that toxinology will be a completely separate discipline from toxicology, which, with the advent of environmental toxicology, agricultural toxicology, forensic toxicology, occupational toxicology and analytical toxicology has, for all practical purposes, detached itself from traditional pharmacology. I believe the latter separation is proper, although the process is suffering from many growing pains within the academic structure.

Finally, the word *biotoxins* seems to me to be redundant and connotative, since if toxins are naturally-occurring plant or animal poisons they must, of necessity, be biological. Arsenic, of course, is a toxic substance not a toxin. Its study falls within the realm of inorganic substances and it is a subject within toxicology. Saxitoxin, on the other hand, is a poison of "life" or "organic life". To add the prefix "bio" seems redundant. Biochemistry is the study of the chemistry *of* life; biodynamics is the study of the actions *of* living organisms; biolytic is the destruction *of* life, etc.

I am also not sure that one can defend biotoxins on the basis that it means toxins that affect biological systems, since there does not appear to be an accepted precedence for this usage. Examples can be cited, I believe, but all such examples appear archaic, limited to historical usage, or nounce-words. I feel the word *toxin*, to imply a venom or poison, or a part of a complete poison or venom, is consistent with present usage (Russell, 1980b) and the science related to the study of these poisons is best termed toxinology.

## C.  *General Chemistry and Pharmacology of Marine Poisons*

While marine toxins as a whole are far more varied in their chemical composition than those from terrestrial animals, there is some degree of component consistence within a particular genus or species of each group.

However, there are some notable exceptions, and there appears to be a far greater qualitative diversity for a single marine toxin than for a particular toxin of a terrestrial animal. Some organisms, such as the clams and mussels, may be toxic only during one period, or a particular period, or in one place, and not elsewhere or during a subsequent season or year; while the toxicity in tetraodons varies with the species of fish, the organs studied, and other factors. Toxicity in ciguateric fishes is, at the present time and for all practical purposes, almost unpredictable with respect to the species involved, location, time of year and other factors.

In the venomous marine animals there are some chemical relationships between the various toxins derived at the oral pole. In general, these are used in an offensive stature; while those derived from the aboral pole, and used in defence, may also have similarities. As one might expect, those protein venoms delivered from the oral pole tend to be higher in enzyme content, while those delivered from the aboral pole tend to contain greater amounts of pain-producing substances, although there are notable exceptions.

Some marine toxins are proteins of low molecular weight, while others are of obvious high molecular weight. Some marine venoms or poisons are composed of lipids, amines, quinones, quaternary ammonium compounds, alkaloids, guanidine bases, phenols, steroids, mucopolysaccharides, or halogenated compounds. The fish venoms are unstable but most of the other toxins, including the fish poisons, are relatively stable, particularly in the dried or lyophilized form. In some marine organisms there are several toxins present, and in some instances, two organisms are necessary to produce one toxin, as seems the case with the blue-green alga *Lyngbya majuscula* Gomont and a dinoflagellate. Finally, it is known that the venom of one species or genus within one phylum may be similar or even identical to that found in an animal of an entirely different phylum. The newt poison, tarichatoxin, and the pufferfish poison, tetrodotoxin, are one and the same.

As would be expected, the pharmacological or toxicological properties of marine toxins vary as remarkably as do their chemical activities. Some marine toxins provoke rather simple effects, such as transient vasoconstriction or dilatation, pain, or localized erythema, while others produce more complex responses, such as parasympathetic dysfunction or multiple concomitant changes in cardiovascular or blood dynamics. In addition to the effects of the separate and combined activities of the various fractions and the metabolites formed by their interactions, there may be a complication by the response of the envenomated organism. The release of histamine, serotonin and perhaps other tissue-active substances from body cells following envenomation may not only complicate a poisoning but may produce more serious consequences than the toxin itself. And there is no doubt that in the evolution of marine toxins, as in snake and other terrestrial venoms (Russell, 1980b), synergistic and possibly antagonistic reactions may occur as the result of interactions between individual venom components.

It seems wise to keep in mind that there is no piece of experimental evidence that demonstrates that the total chemical or pharmacological effect of a whole venom is equal to the sum of the properties of the individual fractions or functions, nor from a philosophical posture does such a conclusion seem plausible. It is one of the unfortunate facts in the study of the chemistry and pharmacology of all the naturally-occurring poisons that the structure and design are most easily investigated by taking the poison apart. This has two shortcomings: it means that a destructive process must be substituted for a constructive, progressive and integrative one; and secondly, the essential quality of the whole toxin may be destroyed before one has made a suitable acquaintance with it. Often the process of examination becomes so exacting that the end is lost sight of in the preoccupation with the means, so much so that in some cases the means becomes substituted for the end.

No comprehensive classification for marine toxins now exists. Our knowledge of the chemical and pharmacological properties of these complex toxins is not broad enough nor consistent enough, at the present time, to permit the adoption of a single working classification. The use of such widely accepted diversions as "neurotoxins," "haemotoxins," "cardiotoxins," "myotoxins," "cytotoxins," for one, is unfortunate, for those terms have usually been applied to a whole venom or poison and most toxins are complex mixtures having several or many biological activities. Even when individual fractions are considered, it has also been shown that neurotoxins can, and do, have cardiotoxic or haemotoxic activities, or both; cardiotoxins can have neurotoxic or haemotoxic activities, or both; and haemotoxins can have other activities. Until the individual fractions of venoms have been isolated and studied separately and in combination, we need to exercise extreme care in systematizing data which are based partly on biochemical studies, partly on biological assay methods, partly on clinical observations and partly on intuitive hunches.

It is true that there are some venom or poison components that may well have an effect at some specific site or membrane in the nervous system, but the basis for calling this fraction a neurotoxin is usually founded on its action on the particular preparation the investigator has selected; and rarely are the other possible pharmacological properties or tissue modulations studied in detail. Our techniques are now sufficiently advanced to explore more specific pharmacological mechanisms at cell membranes, on cells, or at definitive tissue sites. We may find that the basic mechanisms involved are not site-specific but mechanism-specific and on several or many different kinds of tissues, whether the mechanism be ion or protein fluxes, or other basic chemical or electrical phenomena. I have dealt with these matters in a more specific way elsewhere (Russell, 1971a,b; 1980a).

Probably, the most difficult problem faced by the toxinologist is that of proposing a simple, applicable and reproducible biological assay from which data can be transferred and applied in a meaningful way. Unfortunately, and all too frequently, data from a definitive tissue preparation are applied directly to the intact animal, including man. This has resulted in grave errors in clinical judgment (Russell, 1975; 1980b). Great care should be taken in interpreting data from isolated tissue preparations when such information is to be transliterated. The scientist does not have the privilege of speculating too distantly from his data and must exercise extreme care when such data may need to be applied to clinical problems.

This task of providing a suitable bioassay for marine toxinology has been a formidable one. How does one interpret the $LD_{50}$ in haemolytic terms, or vice versa? Of what significance is the $LD_{50}$ in a defensive (repelling) stature? What is the significance of the intraperitoneal or intravenous $LD_{50}$ for toxins in which toxicity develops through ingestion? What is a "mouse unit"? What is the application of a neuromuscular blocking phenomenon in the squid axon to patients who die in hypovolemic shock? Needless to say, all pharmacological data must be interpreted with care and transliteration must always be viewed with suspect. Also, comparisons in data, such as the "haemolytic index", for different animals is questionably valid, unless one considers all the different pharmacological parameters (Russell, 1980a).

In 1965, in preparation for the first international meeting of the International Society on Toxinology, Mr. J. Strassberg and I sent out questionnaires to 100 individuals in 50 countries who were familiar with poisonings by marine animals. From those data we found that fewer than 20 000 cases of poisoning occurred each year following the ingestion of toxic marine animals and probably less than 200 of these were fatal (Russell, 1971a,b). However, in only about 30% of all these cases was the aetiological factor determined. There were less than 80 recorded deaths from pufferfish poisoning, approximately five attributed to ciguateric fishes, three due to paralytic shellfish poisoning, and six reported from eating unknown marine animals under unknown circumstances. However, the study was definitely limited to those areas and individuals who we thought would possess the most significant data and experience but subsequent letters from colleagues indicate that the "less than 200" figure might not be too far off. A more supportable figure would be 150 deaths per year.

With respect to stingings by marine animals, again there are no world-wide compilations. In 1971b, I reported that approximately 5600 persons a year were stung by cnidarians, stingrays, catfishes and the California sculpin along the coasts of North America. In Australia, J. Barnes (personal correspondence, 1965) estimated that almost 5000 persons were stung by cnidarians, stingrays, catfishes and stonefishes each year. J. Morales (personal correspondence, 1965) stated that at least 3000 persons a year in Brazil were stung by venomous marine animals, and incomplete records from fishermen in the North Sea would indicate that during 1958 at least 2000 persons were stung by the weeverfish alone off England, France, Belgium, Holland and Denmark.

> From these and other figures gathered by the author during the past 15 years it would appear that the number of injuries by venomous marine animals must exceed 40,000 a year. Fortunately, the number of deaths, excluding those due to sea snake venom poisoning, is small, probably less than 75 a year (Russell, 1971a,b).

Since 1971, I have gathered additional data on the stingings by venomous marine animals. For instance, Maretić *et al.* (1980) reported that in 1978, 250 000 persons were stung by one species of coelenterate along the northern Adriatic coast of Yugoslavia. Letters to marine stations on the European and African coasts of the Mediterranean indicate that at least 1 000 000 persons are stung by cnidarians each year in that area. Updated records from the United States, Australia, South Africa, Hawaii, Tahiti and the Philippines would indicate that the number of stingings throughout the world would average 1 500 000 a year. As near as I am able to determine, the number of deaths from such stingings, excluding those caused by secondary infection, is less than 50 a year.

## II. Protista

Protista is a kingdom taxon suggested by Ernst Heinrich Haeckel as the not-quite-plant/not-quite-animal ancestors of all living things. The name he coined amounted to the claim that in this taxon we have the "very first" living creatures (Jennings and Aker, 1970). Although that hypothesis might be questioned, the word has survived and it is now used to denote those plant and animal unicellular organisms which have remained at the simple cellular grade of construction. Among the protista are the various protozoans, algae, diatoms, bacteria, yeasts and fungi. In this contribution, a number of protistans found in fresh water have been included because of the similarities of their toxins to those found in marine organisms.

The marine protista are widely distributed throughout neritic waters and in the high seas from the polar oceans to the tropics. There are at least 80 species that are known to be toxic to man and other animals (Table 1). Most of these toxic protista are of the order Dinoflagellata, of which there are more than 1200 species. Dinoflagellates are eukaryotic organisms, principally free-living, motile unicells, although some are coccoid, filamentous or sac-like parasites.

Blooms of protista sometimes occur and result in the phenomenon frequently referred to as "red tide," or "red water". However, the bloom may appear yellowish, brownish, greenish, bluish, or even milky in colour,

**CYANOPHYTA**
   Order:  Chroococales
   Family: Chroococcaceae
     *Microcystis aeruginosa* Kutzing        Toxic to domestic and laboratory
      = *Anacystis cyanea* (Kutz)        animals and fish
      Dr. & Dail
     *Microcystis toxica* Stephens        Toxic to domestic animals and fish

   Order:  Oscillatoriales
   Family: Nostocaceae
     *Anabaena flos-aquae* (Lyngbye)        Toxic to domestic and laboratory
      Brebisson        animals; dermatitis; G.I. upset
     *Aphanizomenon flos-aquae* (L.) Ralfs    Toxic to fish
     *Nostoc spumigena* (Mert.) Drouet     Toxic to terrestrial animals
      = *Nodularia spumigena* (Mert.)
      Bornet & Flahult

   Family: Oscillatoriaceae
     *Lyngbya majuscula* Gomont        Dermatitis
     *Oscillatoria nigroviridis* Thwaites
     *Schizothrix calcicola* (Agardh)
      Gomont
     *Trichodesmium erythreum* Ehrenberg

   Family: Rivulariaceae
     *Gloeotrichia echinulata* (Smith)     Toxic to terrestrial and laboratory
      Richt.        animals

**CHLOROPHYTA or CHLOROPHYCOPHYTA**
   Order:  Caulerpales
   Family: Caulerpaceae
     *Caulerpa* sp.        Marine kills

   Order:  Chlorellales
   Family: Chlorellaceae
     *Chlorella pyrenoidosa* Chick      Toxic to terrestrial and laboratory
       animals

   Family: Scenedesmaceae
     *Scenedesmus obliquus* (Turp.) Kutz.

**PHAEOPHYTA or PHAEOPHYCOPHYTA**
   Order:  Fucales
   Family: Sargassaceae
     *Turbinaria ornata* J. Aq.        G.I. distress

**CHRYSOPHYTA or CHRYSOPHYCOPHYTA**
   Class:   Bacillariophyceae
   Order:  Pennales
   Family: Fragilariaceae

TABLE I. (CONTINUED)

*Fragilaria striatula* Lyngb.

Class: Chloromonadophyceae
Order: Chloromonadales
  *Hornellia marina* Subrahmanyan             Marine kills
  *Hemieutreptia antiqua* Hada                 Marine kills

Class: Chrysophyceae
Order: Ochromonadales
  *Ochromonas danica* Pringsheim          Lab. kills
  *Ochromonas malhamensis* Pringsheim    Lab. kills

Class: Prymnesiophyceae
Order: Prymnesiales
  *Prymnesium parvum* N. Carter         Fish kills, toxin

PYRRHOPHYTA or PYRRHOPHYCOPHYTA
Class: Desmophyceae
Order: Prorocentrales
Family: Prorocentraceae
  *Prorocentrum* sp. (= *Exuvialla* sp.)    Marine kills; G. I. distress
  *Prorocentrum balticum* (Lohmann)     PSP
    Loeblich
  *Prorocentrum cordatum* (Ostenfeld)   Marine kills
    Dodge
  *Prorocentrum linna* (Ehrenberg)       Toxic
  *Prorocentrum minimum* var.           Venerupin
    *mariae-lebouriae* (Parke &
    Ballantine) Faust

Class: Dinophyceae
Order: Dinophysiales
Family: Dinophysiaceae
  *Dinophysis fortii* Pavillard           G.I. distress

Order: Gymnodiniales
Family: Glenodiniaceae
  *Glenodinium folaceum* Stein          Toxic
  *Glenodinium rubrum* Whitelegge      Fish kills

Family: Gymnodineaceae
  *Amphidinium* sp.                      Fish kills
  *Amphidinium carterae* Hulbert       Fish kills
  *Amphidinium rhynchocephalum*       Fish kills
    Anissimowa
  *Cochlodinium* sp.                   Fish kills
  *Cochlodinium catenatum* Okamura   Fish kills
  *Cochlodinium heterolobatum* Silva   Marine kills
  *Gymnodinium* sp.                  PSP; G.I. and respiratory irritant
  *Gymnodinium* type '65           Toxic
    (= *nagdsaki* Iizuka & Nakashima)
  *Gymnodinium flavum* Kofoid & Swezy  Fish kills
  *Gymnodinium galatheanum* Braarud  Marine kills
  *Gymnodinium mikomoto* Miyaki &    Marine kills
    Kominani

| | |
|---|---|
| *Gymnodinium sanguineum* K. Hirasaka (= *splendens* Lebour) | Marine kills |
| *Gymnodinium veneficum* Ballantine | PSP |
| *Gyrodinium aureolum* Hulburt | Fish kills |
| *Ptychodiscus brevis* (Davis) Steidinger (= *Gymnodinium breve* Davis) | PSP; marine kills |
| Family: Polykrikaceae | |
| *Polykrikos schwartzi* Butschli | Toxic |

Order: Noctilucales
Family: Noctilucidae

| | |
|---|---|
| *Noctiluca miliaris* Suriray | Marine kills |
| *Noctiluca scintillans* (Macartney) Ehrenberg | Marine kills |

Order: Peridiniales
Family: Ceratiaceae

| | |
|---|---|
| *Ceratium tripos* (O. F. Müller) Nitzsch | Fish kills |

Family: Gonyaulaceae

| | |
|---|---|
| *Gonyaulax* sp. | PSP |
| *Gonyaulax monilata* Howell | Marine kills |
| *Gonyaulax polyedra* Stein | Marine kills |
| *Gonyaulax polygramma* Stein | Marine kills |
| *Protogonyaulax* (= *Gonyaulax*) *acatenella* (Whedon & Kofoid) Taylor | PSP |
| *Protogonyaulax* (= *Gonyaulax*) *catenella* (Whedon & Kofoid) Taylor | PSP |
| *Protogonyaulax* (= *Gonyaulax*) *cohorticula* (Balech) Taylor | PSP |
| *Protogonyaulax* (= *Gonyaulax*) *dimorpha* (Biecheler) Taylor | PSP |
| *Protogonyaulax* (= *Gonyaulax*) *fratercula* (Balech) Taylor | PSP |
| *Protogonyaulax phoneus* (Woloszynska & Conrad) Taylor = *Pyrodinium phoneus* | |
| *Protogonyaulax tamarensis* (Lebour) Taylor = *Gonyaulax excavata* (Braarud) Balech | PSP; marine kills |
| *Pyrodinium bahamense* Plate | PSP |

Family: Peridiniaceae

| | |
|---|---|
| *Gambierdiscus toxicus* Adachi & Fukuyo (= "GDT" Taylor) | Ciguatoxin, Maitoxin |
| *Heterocapsa triquetra* (Ehrenberg) Stein | Marine kills |
| *Peridinium polonicum* | Marine kills; laboratory animal kills |
| *Peridinium trochoideum* (Stein) Lemm. | Marine kills; laboratory animal kills |

depending on the organism involved, and other factors (Russell, 1965a,b). Such blooms usually become visible when 20 000 or more of the organisms are present in 1 ml water. However, some blooms may contain 50 000 or more organisms. The red colour in red tides is probably due to peridinin, a xanthophyll.

A number of factors have been suggested as favouring blooms. In 1965a,b I noted these, including changes in weather conditions that bring about upwellings or other alterations in water masses, changes in water temperature and sunlight, changes in those factors which affect water turbulence, transparency, surface illumination, passive sinking of the phytoplankton themselves to depths beyond the photic zone, and the grazing action of the zooplankton population. More recently, a number of other factors have been implicated in blooms and some of the previously suggested aetiological factors have been re-examined.

In his opening address at the Second International Conference on Toxic Dinoflagellate Blooms in Key Biscayne, Florida in 1978, Provasoli noted the progress that had been made in understanding the factors that precipitate dinoflagellate blooms. Of particular interest was the hypothesis advanced by Steidinger (1975) and by Wall (1975) that blooms might be initiated by benthic resting hypnocysts. These authors had reported that viable and durable cysts were present in widely separated areas, and when germinated they produced dinoflagellates. The sexual hypnocysts were found in locations where blooms had occurred the previous year, indicating a cyclic link to succeeding blooms (Anderson and Wall, 1978). Blooms appeared to be temperature-, light-, salinity-, trace ion- and nutrient-dependent, while hydrographic factors, such as storms, winds, tidal mixing, currents, upwellings and dredging were thought to influence resuspension of cysts in sediments.

Provasoli (1978) also called attention to the importance of hydrography in blooms, particularly in the form of phototaxis, differential transport to varying salinities, organism mixing, wind and light, and whether the organism was toxic or nontoxic. Nutrient support has been noted as a very important factor in the production of blooms. Red tides are largely coastal and coastal waters have a particularly high concentration and variety of organic matter, vitamins, trace elements, chelating agents, etc., all of which contribute to dinoflagellate growth. The interested reader will find Provasoli's article to be an excellent overview of our knowledge on dinoflagellate blooms and the text of the conference (Taylor and Seliger, 1979) a valuable reference work.

When excessive numbers of these unicellular organisms collect, there may be a mass mortality of fishes or other marine animals in the area (Carrilo, 1892; Kofoid, 1911; Hornell, 1917; Nightingale, 1936; Galsoff, 1948; Connell and Cross, 1950; Fish and Cobb, 1954; Smith, 1954; Brongersma-

Sanders, 1957; Grindley and Taylor, 1962; White, 1977; National Marine Fisheries Service, 1977; Bodeanu and Usurelu, 1979; De Mendiola, 1979). These mass mortalities can be due to a toxin produced by the protista. However, oxygen depletion in the water, due either to the number of plankton present or to the release of decay products by these organisms and the dying animals may be as devastating. It should be noted that many of the recorded fish and other marine animal kills have been associated with non-toxic protista.

In the case of benthic animal mortality, one contributing cause might be the nocturnal descent of dinoflagellate blooms, which are known to form mats interspersed with mucoid strands, which could either suffocate bottom-dwelling invertebrates or impede their movements and survival (Tangen, 1979). Whatever factors are involved, marine animal kills are certainly associated with oxygen depletion, and to other factors related to the anoxia.

With respect to humans, Halstead (1965) notes that there were 222 recorded deaths from paralytic shellfish poisoning between 1689 and 1962. Prakash et al. (1971) note 80 reported cases around the Bay of Fundy from 1889 to 1970, with three deaths; and 107 cases in the St. Lawrence area, with 21 deaths during the same period. Global figures are not available but these may be close to 2000 cases of paralytic shellfish poisoning each year. Halstead believes the number of cases and the geographical spread is on the increase. This certainly seems reasonable from a review of the 1980–1982 literature.

## A. *Paralytic Shellfish Poisoning*

Paralytic shellfish poison (PSP), variously known as "saxitoxin," "*Gonyaulax* toxin," "dinoflagellate poison," "mussel or clam poison," or "mytilotoxin" is a toxin or group of toxins found in certain molluscs, arthropods, echinoderms and some other marine animals which have ingested toxic protistan and have become "poisonous". Paralytic shellfish poisoning through the food-chain is well known in animals and man. The problem has been reviewed by Russell (1965a), Halstead (1965), Schantz (1971), Shimizu (1978) and Hashimoto (1979).

The principal areas in which toxic protista have been found and are a source of public health concern are along the California coast north to British Columbia, south-eastern Alaska and over the Aleutian Islands, and Japan; on the Atlantic coast of North America, Maine, the Maritime Provinces and the Gulf of Mexico; in Europe off Scotland, England, Ireland, Spain, France, Belgium, Holland, Germany, Denmark and Norway; and off New Zealand, Australia, Papua, New Guinea, Malaysia, South Africa, Venezuela, Peru and Chile.

The relationship between blooms of plankton and shellfish poisoning was perhaps first noted by Lamouroux (cited by Chevallier and Duchesne, 1851), who observed that during certain seasons of the year the sea appeared as a yellowish "foam" and that this foam was probably responsible for the poisonous properties of the shellfish. However, most early workers attributed the poisoning to other causes: copper salts, putrefactive processes, diseases of the shellfish, a "virus", other marine organisms, contaminated water, industrial wastes or bacterial pathogens.

In 1888, Lindner also suggested a food-chain relationship for shellfish poisoning, and subsequently this hypothesis received more favourable consideration. In 1937, Sommer and Meyer published the results of their intensive investigation of the problem of paralytic shellfish poisoning. They demonstrated a direct relationship between the number of *Gonyaulax catenella* Whedon et Kofoid in the sea water and the degree of toxicity in the mussel, *Mytilus californianus* Conrad. These workers also established methods for extracting and assaying the poison and suggested an experimental and clinical approach to the problem that has served as a guide for many subsequent investigations.

The amount of the poison in the shellfish or other organism is dependent upon the number of toxic protistan filtered by the host animal. Off California, mussels become dangerous for human consumption when 200/ml or more protista are found in the coastal waters. As the count rises the mussels become more toxic, and as the count falls they become less toxic. Within a week or two in the absence of the toxic protistan, the mussels become relatively free of the toxin.

The crab *Zosimus aeneus* from Amami-Oshima Island was found by Hashimoto (1979) to contain saxitoxin but there was no evidence of either a protistan bloom in the area or of toxic crustaceans, and the highest level of the toxin was found in the exoskeleton. Shimizu (1978) has suggested that "symbiotic dinoflagellates reside in the animal bodies and constantly produce toxins", making the crabs toxic over long periods of time. This mechanism, however, is open to question.

Paralytic shellfish poisoning has been studied by extracting the toxin(s) from shellfish, dinoflagellates secured from natural blooms and, more recently, from laboratory cultures. It has been demonstrated that PSP can be obtained from all three sources in like form. Burke *et al.* (1960), Schantz *et al.* (1966) and Proctor *et al.* (1975) have grown *Gonyaulax catenella* in axenic cultures in cell densities equal to those occurring during natural blooms, and Schantz (1960) showed that the chromatographic properties of the toxin from the cultured organisms appear identical to that found in natural blooms and mussels. Although it was once suspected that the toxin was formed by a bacterium with the protistan, culturing of dinoflagellates free of bacteria has demonstrated that the toxin is not a symbiotic effect of bacteria.

## B. *Chemistry and Pharmacology*

In the earlier works, PSP was considered as a single poison but it must now be thought of as a complex of toxins. There is little doubt the earlier isolations were impure and that some of their chemical and pharmacologic properties can be attributed to these impurities. In the case of the toxins from *Gymnodinium breve*, for example, there now appear to be several, if not a number of toxins, and at this point in time it has not been established that these toxins reflect differences related to the techniques employed in the isolation procedures or are the result of distinct differences in the toxin(s).

The impetus for the initiation of studies on the chemistry of shellfish poison was a mass poisoning at Wilhelmshaven, Germany in 1885. Following the poisoning, Salkowski (1885) prepared four alcoholic extracts from mussel tissues, concentrated them by evaporation and then reconstituted them with water. When the reconstituted product was forced into alcohol, a viscous precipitate formed. The filtrate from this precipitation contained the poison. Brieger (1888) isolated a substance he called "mytilotoxin", which produced effects similar to those provoked following the eating of toxic mussels. Richet (1907) isolated a substance which caused changes in animals not unlike those described for one of Brieger's toxic fractions. Since these signs were similar to those he had previously noted following poisoning with a toxin ("congestine") from a sea anemone, he called the new poison "mytilocongestine".

In 1922, Ackermann identified a number of bases in extracts from mussel tissues, including adenine, arginine, betain, neosin, methylpyridylammonium hydroxide and crangonine, none of which, however, had the properties of PSP. Partial purification of the toxin from California mussels was obtained by Müller (1935), who used permutit as an absorbent, eluted with saturated potassium chloride, and separated the poison by extraction of the residue from evaporation with methanol. Sommer *et al.* (1948) obtained the toxin following passage of extracts through charcoal, removing the lipid impurities with ether, passing the residue through sodium permutit and then treating with ethanol. On evaporation they obtained a product with a lethality of 6–12 $\mu$g per mouse unit.

Schantz *et al.* (1957) obtained high yields of the toxin from California mussels and Alaska butter clams using chromatography on Amberlite XE-64 prior to chromatography on acid-washed alumina. Mold *et al.* (1957) found that distribution of the toxin in a solvent system of *n*-butanol, ethanol, 0·1 M aqueous potassium carbonate and *a*-ethyl caproic acid in a volume ratio of 146 : 49 : 200 : 5, with the aqueous layer adjusted to pH 8, resulted in a separation of the poison into two components, one of which was slightly more toxic than the other. At that time it was suggested that the poison existed in two tautomeric forms because upon standing in acid solution each of the components equilibrated to form the same mixture. Subsequently, Schantz (1963) showed that both clam and mussel toxins were basic in nature, forming salts with mineral acids. I have summarized the known chemical properties of *Gonyaulax* and mussel poisons as of 1967 elsewhere (Russell, 1971b). In 1967, I presented the structure for saxitoxin that was suggested by Rapoport (personal communication, 1967):

$$\begin{array}{c} H \\ | \\ H-C-O-C \overset{\displaystyle O}{\underset{\displaystyle NH_2}{\diagdown}} \end{array}$$

In 1971, Wong *et al.* of the Rapoport group presented evidence for the perhydropurine skeleton of saxitoxin by degrading it to aminopurine derivatives. Their final structure was:

Further studies with a thoroughly dried sample gave $C_{10}H_{15}N_7O_3 \cdot 2HCl$ rather than $C_{10}H_{17}N_7O_4 \cdot 2HCl$. Schantz *et al.* (1975) finally succeeded in establishing a crystalline di-*p*-bromobenzene sulfonate of saxitoxin and X-ray diffraction studies gave the structure which included the absolute configuration:

Bordner *et al.* (1975), working independently, completed X-ray diffraction of a crystalline $C_{12}H_{21}H_7O_4 \cdot H_2O$ obtained by treating saxitoxin dihydrochloride in an ethanolic solution and confirmed the configuration as a stable hemiketal.

In 1975, Shimizu *et al.* found that in the dinoflagellate *Gonyaulax tamarensis* in addition to saxitoxin, there were several other toxins which differed from saxitoxin only in their weak binding ability on carboxylate resins. Further studies on organisms obtained from red tides along the New England coast resulted in the isolation of two new toxins, "gonyautoxin II" ($GTX_2$) and "gonyautoxin III" ($GTX_3$). Like saxitoxin, these new toxins lose their toxicity in high pH solutions. Both $GTX_2$ and $GTX_3$ were found to be highly hygroscopic. Their molecular weights and molecular formulae have not yet been established. Their structures are shown below. $GTX_2$ is a proposed 11-hydroxysaxitoxin, while the isomeric $GTX_3$ is considered a product of emolization.

Another toxin, neosaxitoxin, a 1-hydrosaxitoxin, has been isolated from G. tamarensis and exhibits chemical properties similar to saxitoxin, with one notable exception being that of its enhanced absorption at $1770 \text{ cm}^{-1}$, which is attributed to a carbonyl function (Shimizu, 1978).

While a number of pharmacological and toxicological studies on shellfish poisons were carried out before the turn of this century, it was not until Meyer et al. (1928), Prinzmetal et al. (1932) and Sommer and Meyer (1937) that the more definitive work was reported. Prinzmetal et al. (1932) showed that the poison from the mussel Mytilus californianus was slowly absorbed from the gastrointestinal tract and rapidly excreted by the kidneys. It was said to depress respiration, the cardioinhibitory and vasomotor centres and conduction in the myocardium. Kellaway (1935a,b) suggested that the toxin had a direct effect on both the central nervous system, particularly the respiratory and cardiovascular centres, and the peripheral nervous system, particularly the neuromuscular junction and sensory nerve endings. The poison caused a rapid fall in systemic arterial pressure and a slowing of respiration. The latter he attributed to the central effects of the toxin.

Sommer and Meyer (1937) found that 3000 Gonyaulax weighed 100 μg (wet weight) and that this number yielded 15 μg of the dry extract, which in turn gave 1 μg of pure poison, or 1 mouse unit. A mouse unit, or average lethal dose, was defined as the amount of toxin that would kill a 20 g mouse in 15 min (Prinzmetal et al., 1932; Sommer and Meyer, 1937). Thus, the amount of toxin contained in a single Gonyaulax was taken as 1/3000 of a mouse unit. Subsequently, various testing methods and assays were studied by Medcof et al. (1947), Meyer (1953) and McFarren et al. (1956). The latest standards are those set by the Association of Official Analytical Chemists (A.O.A.C.) (1975). McFarren et al. (1956) found the oral $LD_{50}$/kg body weight to vary considerably with the animal used and with its strain and weight. Their figures would indicate that the human is twice as susceptible to the poison as the dog and approximately four times more susceptible than the mouse.

During the 1950s, a Canadian–United States Conference on Shellfish Toxicology adopted a bioassay based on the use of the purified toxin isolated by Schantz et al. (1958). Studies utilizing the methods outlined by the Conference indicated that the intraperitoneal minimal lethal dose of the toxin for the mouse was approximately 9·0 μg/kg body weight. The intravenous minimal lethal dose for the rabbit was 3·0–4·0 μg/kg body weight, while the minimal lethal oral dose for man was thought to be between 1·0 and 4·0 mg. Wiberg and Stephenson (1960) demonstrated that the $LD_{50}$ of the then purified toxin in mice was:

| | |
|---|---|
| Oral route | 263 (251–267) μg/kg |
| Intravenous route | 3·4 (3·2–3·6) μg/kg |
| Intraperitoneal route | 10·0 (9·7–10·5) μg/kg |

More recently, various figures on the toxic and lethal doses for Man have been presented by various workers and discussed at meetings of the International Society on Toxinology. The figures presented by Prakash et al. (1971) seem consistent with our own calculations; that is, a mild case of poisoning can be caused by ingesting 1 mg of toxin, which might be the amount found in 1–5 poisonous mussels or clams weighing about 150 g each. A moderate case of poisoning can be caused by ingesting 2 mg of the poison, while a serious poisoning would be caused by 3 mg. One would expect that 4 mg of the toxin would be lethal to a human being if vigorous treatment was not instituted.

The various standards for measuring shellfish toxicity have several short-comings. The currently employed mouse assay, as suggested by the A.O.A.C. (Association of Official Analytical Chemists, 1975), is a reasonable method but harbours some failings. A number of assays have been proposed to circumvent these deficiencies. For example, an immunochemical technique was suggested by Johnson et al. (1964), while Bates and Rapoport (1975) described an analysis based on oxidation of saxitoxin to a fluorescent derivative. Spectrophotometric analysis has been proposed (Gershey et al., 1977), and a unique cockroach bioassay was described by Clemons et al. (1980a,b). One of the most promising assays incorporates flow cytometric analysis of cellular saxitoxin, dependent on mithramycin fluorescent staining (Yentsch, 1981). With the advent of the enzyme-linked immunosorbent assay (ELISA) and radioimmunology, new and improved techniques for determining toxicities, hopefully multiple toxicities, should appear within the next few years.

It was found that saxitoxin had a marked effect on peripheral nerve and skeletal muscle in the frog. The "curare-like" action was attributed to some mechanism which prevented the muscle from responding to acetylcholine

(Fingerman *et al.*, 1953). Bolton *et al.* (1959) obtained somewhat similar results; they demonstrated a progressive diminution in the amplitude of the end plate potential of the frog nerve-muscle preparation exposed to the toxin.

Murtha (1960) also noted that the toxin depressed mammalian phrenic nerve potentials, suppressed the indirectly-elicited contractions of the diaphragm and often reduced the directly-stimulated contractions. It was concluded that the effect of the poison is greater on reflex transmission than on the nerve. Woodward (1955) presented oral lethal doses for eight animal species (Table II). The table and its significance have been discussed by McFarren *et al.* (1960).

TABLE II. ORAL $LD_{50}$ PER KILOGRAM IN MOUSE UNITS FOR CRUDE ACID EXTRACT OF TOXIC CLAMS (FROM WOODWARD, 1955)

| Animal | $LD_{50}$ |
|---|---|
| Mice | 2100 |
| Rats | 1060 |
| Monkeys | 2000–4000 |
| Cats | 1400 |
| Rabbits | 1000 |
| Dogs | 1000 (approximately) |
| Guinea-pigs | 640 |
| Pigeons | 500 |

According to Pepler and Loubser (1960), the toxin had a very marked specific acetylcholinesterase inhibitory effect, similar to that of the organophosphorus compounds. This point, however, was open to considerable question. Schantz (1960) indicated that the contraction of isolated muscle fibres in the presence of ATP and magnesium ions was not inhibited by the poison, nor did the toxin alter the rate of oxygen consumption in the respiring diaphragm of the mouse.

At the Symposium on the Biochemistry and Pharmacology of Compounds Derived from Marine Organisms in 1960, Murtha presented evidence indicating that the toxin had a direct effect on the heart and its conduction system. He noted that it produced changes which ranged from a slight decrease in heart rate and contractile force, with simple P–R interval prolongation or S–T segment changes, to severe bradycardia and bundle-branch block, or complete cardiac failure. He also demonstrated that it provoked a prompt but reversible depression in the contractility of isolated cat papillary muscle. In those cases where spontaneous contractures were abolished, Murtha found that the muscle still responded to electrical stimulation, although contractile force was reduced approximately 50%.

In both intact and partially eviscerated mammals he observed a precipitous fall in systemic arterial pressure following injection of the toxin, indicating that the mechanism proposed by Kellaway (1935a,b) (changes in the splenic circulation) was not responsible for the cardiovascular crisis. In vagotomized dogs, cardiac contractile force decreased 50% within the first minute following injection of the poison. There was a concomitant precipitous fall in systemic arterial pressure. In cervical cord-sectioned, bilaterally vagotomized mammals the immediate precipitous fall in arterial blood pressure was not seen, although some decrease in pressure subsequently occurred. The toxin did not produce vasodilation in the vessels of the mammalian leg or kidney, nor did it affect the rate of blood flow in the isolated rabbit ear.

These various studies on the action of paralytic shellfish poison on the cardiovascular system indicated that the toxin has a direct effect on the heart, an effect which is in part responsible for the cardiovascular crisis; and while the poison may produce changes in the peripheral vascular system, these changes are probably not of sufficient magnitude to precipitate deleterious alterations in the systemic arterial blood pressure. It also appears that a part of the cardiovascular crisis is in some manner concerned with the direct action of the poison on the central nervous system, although the experiments to that date did not exclude the possibility that cerebral anoxia secondary to cardiac-centred vascular failure may be a factor. Murtha (1960) suggested that the central nervous system effect may be mediated through the spinal cord.

In comparing the pharmacological effects of tetrodotoxin and saxitoxin, Cheymol (1965) stated that the former was more active than the latter on the nerve but the latter was slightly more active on the neuromuscular preparation, even though its effects were more easily reversible.

At the First International Symposium on Animal Toxins in Atlantic City, Kao (1967) demonstrated that the toxin blocks action potentials in nerves and muscles by preventing, in a very specific manner, an increase in the ionic permeability which is normally associated with the inward flow of sodium. It appeared to do this without altering potassium or chloride conductances. At the same meeting, Evans (1967) showed that in cats, mussel poison blocks transmission between the peripheral nerves and the spinal roots. The large myelinated sensory fibres are blocked by intravenous doses of 4·5–13 μg/kg while the large motor fibres are not blocked until this dose is increased by approximately 30–40%.

In a subsequent paper, Evans (1968) observed that when dilute solutions of saxitoxin were applied locally to thin peripheral nerve branches in cats, conduction was not blocked. However, conduction was blocked in dorsal and ventral spinal root fibres following the topical application of far smaller concentrations. He suggests that one of the layers in the connective tissue sheath of peripheral nerve is impermeable to saxitoxin, while the leptomeninges covering the spinal roots are either deficient in or lack this layer.

There is some disagreement as to the chemical nature of the toxins from *Gymnodinium* spp. and, in particular, *G. breve*. This is due, in part, to the relative impurity of some of the earlier products, although it now appears that the chief differences are more directly related to techniques, that is, to the differences in methods employed by the various investigators in isolating the final compounds. Shimizu's (1978) review handles these discrepancies well and the reader is referred to his work for a better understanding of the various *G. breve* toxins.

There appear to be at least three different toxins in *G. breve*. In 1957, Ballantine and Abbott obtained a toxin from *G. veneficum* Ballantine that was soluble in water and dilute alcohol, insoluble in ether and chloroform, nondialyzable and could be decomposed by hot alkali. Starr (1958) showed that *G. breve* toxin was similar to that from *G. veneficum*. Paster and Abbott (1969) extracted a haemolysin with chloroform-methanol (2 : 1) from axenic cultures of *Gymnodinium*. In 1970, Martin and Chatterjee isolated two toxins from laboratory blooms of *G. breve* by acidifying the sea water and shaking it with chloroform. The first toxin (I) was found in the emulsion formed at the interface. It was separated and subsequently named "gymnodin" (Doig and Martin, 1973). The second toxin (II) was isolated by passing the chloroform extract through a silica gel column, washing with methylene chloride, eluting the residual material with ethanol and treating the product with charcoal. Its molecular weight was estimated to be 650 and its empirical formula $C_{90}H_{162}O_{54}P$. A toxin similar to substance II was isolated by exposing etheral extracts of *G. breve* cultures to thin layer chromatography on silica gel. Its elemental composition was C, 63·2; H, 8·9; O, 26·4; P, 1·2%, and a molecular weight of 468 was obtained by osmometry (Trieff *et al.*, 1972).

It was found that gymnodin did not cause haemolysis as did prymnesin. Martin and Padilla (1974) considered the possibility that since gymnodin was not haemolytically active it might affect potassium transport. They examined potassium flux in erythrocyte suspensions and concluded that the mode of action of the toxin was not dependent upon differences in the relative internal concentrations of sodium or potassium.

In a very well executed group of experiments, Spiegelstein *et al.* (1973) processed cultured cells of *G. breve* by centrifugation, pelleting, resuspension in water, lyophilization, treatment with chloroform methanol and pure methanol, evaporation, further chloroform-methanol-water treatment, centrifugation, drying and evaporation. The product, a green powder, was applied to a Sephadex $LD_{20}$ column and the fractions collected. Fractions 6–16 proved lethal to fish, frogs and mice, while fractions 33–43 showed haemolytic activity *in vitro*. Tubes 6–16 were dissolved in methanol to which petroleum ether was added. The resulting two phases were separated by centrifugation and the lower phase further processed on Biorad (10–20 μ) and a yellow toxic fraction, "GT", was evaporated to dryness.

Most of the neurotoxic activity was found in this GT. The GT was subjected to thin-layer chromatography and the final products, including the neurotoxic components "$T_1$ and $T_2$", were assayed. It was found that the intravenous $LD_{50}$ in mice was approximately 0·15 mg/kg body weight and that the mice died in 1–3 min. $T_2$ also had a low $LD_{50}$, while that for $T_1$ was five times higher. Frogs were considerably more resistant to the toxins, while fish were quite sensitive. Although the authors call the T fractions neurotoxic, which they may certainly be, these authors' careful observations gave rise to the statement that

> these (kicking, jumping and jerking) motions, almost convulsive, occur during the agonal period and reflect the hypotensive crisis and the change in vascular parameters . . . It seems wiser to explain the death . . . as being due to cardiovascular failure rather than direct effect on the nerve muscle junction.

The error these authors are alluding to is the habit of calling a toxin a neurotoxin on the basis of gross observations during an agonal period of a hypotensive crisis when, among other things, the blood supply to the brain and lungs is reduced. Under such conditions, an animal is quite likely to jump, jerk, tremble, or drag its hind limbs. This should not be considered a neurotoxic effect without further evidence. A classical example of this misconception was once demonstrated by a professor of mine who chopped off the head of a chicken in front of the class and pointing to the axe, said: "This is what is known as a neurotoxin. Now all we must decide is which part of the chicken is having the convulsion".

In the following year, Grunfeld and Spiegelstein (1974) noted that GT toxin exerted its spasmogenic effect on the guinea-pig ileum through stimulation of the post-ganglionic cholinergic nerve fibres.

Alam et al. (1975) isolated three toxins from cultures of G. breve. One of the three, "$T_2$", chromatographically homogeneous, had a molecular formula of $C_{41}H_{59}NO_{10}$ and an estimated molecular weight of 725. However, as Shimizu has pointed out, $T_2$ may be impure or contain several toxins. A further purified fraction of $T_2$, namely GB-2, showed a different u.v. absorption and a molecular weight of 850. GB-2 appears to be a large alkyl ether molecule with a few carbonyl functions (Shimizu, 1978).

Runnegar and Falconer (1975) demonstrated that alkaline extraction of lyophilized G. breve contained a peptide without a free amino group, which on hydrolysis yielded equimolar amounts of L-methionine, L-tyrosine, D-alanine, D-glutamic acid, erythro-β-methylaspartic acid, and methylamine. The mouse $LD_{50}$ of this toxin was 0·056 mg/kg.

Other G. breve toxins have been isolated by Sasner et al. (1972) and Padilla et al. (1974), while the toxicity of additional dinoflagellates has been noted by McLaughlin and Provasoli (1957), Wangersky and Guillard (1960) and Ikawa and Taylor (1973). Kim and Padilla (1977) identified six major fractions in cell cultures of G. breve. Three of these had haemolytic toxicity but none were toxic to fishes. Their isolation technique is shown in Fig. 1.

The most recent purification process for G. breve has involved a rather definitive stepwise procedure described by Baden et al. (1981). In essence, the technique involves extraction procedures, silica gel and thin-layer chromatography, elution, evaporation, precipitation and recrystallization. The technique yields toxins T17 and T34. T34 is further purified through an ethyl acetate/petroleum ether and thin-layer chromatography, which yields a stable crystalline aldehyde. It is believed that T34 is similar if not identical to GB-2 isolated by Shimizu (1978). The purified T34 showed a 2·5-fold increase in lethality to fish and a 1·7-fold increase in lethality to mice than previously reported (Baden *et al.*, 1979). Preliminary studies on the squid giant axon and rat phrenic nerve-diaphragm preparations indicated that T34 acts as a depolarizing agent.

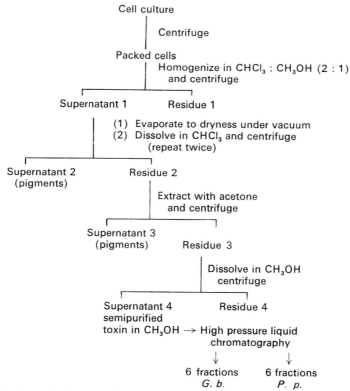

FIG. 1. Isolation technique modified from Kim and Padilla (1977).

The dinoflagellate *Gonyaulax monilata* Howell was first described following a fish-kill bloom off the east coast of Florida (Howell, 1953). Clemons *et al.* (1980b) concentrated a toxin from harvested cells as shown in Fig. 2. The toxin caused death in 1·5–2·5 cm guppies at the 30 $\mu$g/ml level, was lethal to mice at 0·02 mg/kg body weight, had an $LD_{50}$ for roaches of 138 $\mu$g/g and caused 50% haemolysis of human erythrocytes at 7·7 $\mu$l/ml. The organism did not appear to have saxitoxin-like toxins but its cytolytic activities appeared very similar to those of *P. parvum* and *G. breve*. As Clemons points

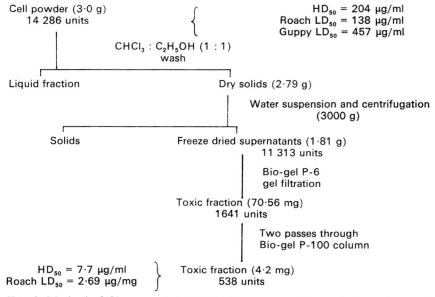

FIG. 2. Method of Clemons *et al.* (1980b) for concentrating toxin from *Gonyaulax monilata*.

out, the toxins of this dinoflagellate, like those from most protista, are only weakly extractable in mild aqueous solvents. Prymnesin and *G. breve* toxins are best extracted with organic solvents, while saxitoxin is extracted with acidified mixtures of alcohol and water. It might be expected that strong cellular retention of the toxins in water affords advantages to the species.

For the most part, the toxic protistans so far discussed have been protozoan and marine in nature. At this point attention will be given to the algae, both marine and freshwater forms, since some of the toxins found in protozoan and algae have quite similar chemical and pharmacological properties. Perhaps as our knowledge increases, the similarities and relationships between these toxins will become more apparent and we may find a common chemical evolution for protistan toxins. The interested reader should consult Collins' (1978) review on algal toxins.

Algae may be either marine or freshwater chlorophyll-bearing organisms, ranging in size from the microscopic unicellular blue-green Cyanophyta of approximately 0·5 μm in diameter to the giant kelps, which may reach more than 30 m in length. All algae are nonvascular. Some algae, such as the cyanophytes, can convert nitrates to nitrites and thus share a role with bacteria in soil reconditioning.

Most marine algae are restricted to certain latitudes and, as a whole, to certain oceanic regions or particular coasts. Some species show a vertical zonation related to light penetration, tide action and water chemistry.

Marine algae may be found to depths of 180 m in the equatorial zone where solar rays range vertically. Drifting algae (phytoplankton) may become entangled among attached algae near or on shore, generally in intertidal zones. These organisms can withstand drying when exposed above the ebb-tide and are physiologically and structurally well adapted for almost any change where moisture is not a problem.

The blue-green algae (Cyanophyta) are non-motile and among the simplest of plants, so well adapted to variable environments that they can be con-sidered ubiquitous. They are found in fresh water, brackish water and salt water, in hot springs and polar icecaps and in terrestrial settings from humid areas to the desert. They are often referred to as "cyanobacteria" because of their resemblance to bacteria in their lack of organized nuclei and manner of cell division. They are not closely related to the other algal groups (Moore, 1981).

Some freshwater and marine blue-green algae are known to be toxic to humans and to many animals (see Table I). Among the animals that have been affected are man's domestic animals, fish, oysters, barnacles, coquians, shrimps, crabs and zooplankton. For instance, Lightner (1978) has shown that blooms of the blue-green alga *Spirulina subsala* give rise to a high incidence of mortality in the blue shrimp *Penaeus stylirostris*. Death is caused by over-whelming necrosis of the epithelial lining of the midgut, dorsal caecum and hindgut gland, leading to haemocytic enteritis. It is believed that under certain conditions the alga produces a weak toxin which attacks the midgut, although the mechanism for the toxicity is not known.

Among the toxic freshwater blue-green algae are *Microcystis aeruginosa*, *Anabaena flos-aquae* and *Aphanizomenon flos-aquae*. The earlier work on *M. aeruginosa* was carried out by Olson (1951), Bishop *et al.* (1959) and others. It was found that the toxic principle was soluble in water and methanol, and insoluble in acetone, ether and benzene. It diffused through collodion, cellophane and animal membranes, and was heat stable at neutral pH. With the mass culturing of toxic strains (Hughes *et al.*, 1958) more definitive and better controlled investigations were initiated. A substance called "FDF" (fast-death factor) was isolated from the alga and differentiated from "SDF" (slow-death factor), which was associated with bacterial action. The purified toxin contained seven amino acid residues, with leucine seeming to play some important role in toxicity. The estimated molecular weights were 1600 and 2600, and the mouse intraperitoneal $LD_{50}$ was 0·46 mg/kg body weight (Bishop *et al.*, 1959). Subsequently, analysis of the toxin hydrolyzate indicated seven more amino acids (making a total of 14) (Rama Murthy and Capindale, 1970) and a mouse intraperitoneal $LD_{100}$ of 0·1 mg/kg body weight.

The principal toxin of *Anabaena flos-aquae* is a water-soluble alkaloid, chemically similar to cocaine and certain tropane alkaloids. It is now known as "anatoxin-a" (ANTX-A) and its configuration has been presented by

Huber (1972). It exhibits mammalian neuromuscular depolarizing properties similar to those of decamethonium bromide, to which, of course, it is not structurally related. It is said to produce rapid toxicity by the oral route (Carmichael *et al.*, 1979). Several new toxic clones have been grown on axenic culture and appear to have different pharmacological properties from those of ANTX-A (Carmichael and Gorham, 1978). These are sometimes called anatoxins B, C and D.

The toxic property of *Aphanizomenon flos-aquae* is an endotoxin and thought to be due to saxitoxin and perhaps two or three unrelated compounds (Collins, 1978). It was once classified as a "very fast death factor" (VDF) because of its rapid death time. The minimal lethal dose in mice is between 0·05–0·10 mg/kg body weight (Gentile, 1971). Again, this end point is of questionable reliability and attempts to compare toxins on this basis alone may not be valid. *Aphanizomenon* toxins block action potentials, suspectedly interfering with sodium transport without affecting potassium-dependent resting potentials (Gentile, 1971). In this respect it has neurotoxic properties similar to saxitoxin and tetrodotoxin but is said to differ in that it also blocks some calcium-dependent action potentials (Fusetani *et al.*, 1976).

There are a number of marine blue-green algae of toxicological importance. Arnold *et al.* (1959) described a contact dermatitis in bathers off Hawaii. They termed the lesions "seaweed dermatitis" and attributed them to a marine algae. Grauer (1959) further described the dermatitis and it soon became known as "swimmers' itch", "surf itch" or "seaweed dermatitis". The disease was attributed to *Lyngbya majuscula* Gomont (now *Microcoleus lyngbyaceus*) (Grauer, 1959; Grauer and Arnold, 1961).

Banner (1959) found that the dried alga incorporated into pellets or a gelatin solution, killed mice when given orally. It should be noted that not all lyngbya species are toxic and even the toxicity of *M. lyngbyaceus* can vary from locale to locale and at different times of the year.

Subsequently, the inflammatory agent of the alga was found to be debromo-aplysiatoxin (Mynderse *et al.*, 1977), a substance which had previously been identified in the digestive gland of the sea hare *Stylocheilus longicauda*.

Debromoaplysiatoxin

Debromoaplysiatoxin was shown to produce a pustular folliculitis in humans and a severe cutaneous inflammatory reaction in the rabbit and hairless mice. The toxin is one of the most potent skin irritants known (Solomon and Stoughton, 1978). It has also been found in an algal mixture, which included *Oscillatoria nigroviridis* and *Schizothrix calcicola*.

In addition to the debromoaplysiatoxin and 19-bromoaplysiatoxin found in the mixture of *O. nigroviridis* and *S. calcicola*, Mynderse and Moore (1978) identified several other toxic components, the most important of which they term "oscillatoxin". In addition to these substances, sterols, volatile constituents, fatty acids, malyngamides, pyrrolic compounds, free amino acids and other constituents have been identified in the blue-green algae but the specific pharmacological activities of these components, as well as of most of the algal toxins are unknown.

An additional vesicating substance has been identified in another shallow-water variety of *M. lyngbyaceus* found off Hawaii. This is an indole alkaloid, "lyngbyatoxin A", having the formula $C_{27}H_{39}N_3O_2$, and a mouse minimal lethal dose of 0·3 mg/kg body weight (Moore, 1977; Cardellina *et al.*, (1979).

Lyngbyatoxin A

It is known to produce escharotic stomatitis if eaten (Sims and VanRilland, 1981). Another blue-green alga, *Trichodesmium* sp., is known as "sea sawdust". It is a red, blue-green alga and is associated with massive fish kills. The nature of its toxin is not known.

Drinking water contaminated with freshwater strains of *Oscillatoria* sp. or *S. calcicola* are known to cause gastrointestinal disturbances in humans (Schwimmer and Schwimmer, 1964; Lippy and Erb, 1976). It has been suggested that the toxins might be lipopolysaccharides (Keleti *et al.*, 1979). In his earlier work on the possible etiological agents responsible for ciguatera poison, Banner (1967) found two lipid-soluble toxins and a water-soluble toxin in *S. calcicola* but they were not characterized, nor did he feel that they were ciguateric. As Moore (1981) has noted, however, the lipophilic toxins may be related to the aplysiatoxins or oscillatoxins.

The golden or yellow-green algae (Chrysophyta) are of particular importance because of the members of *Prymnesium parvum* Carter, *Ochromonas danica*, *O. malhamensis* and *Fragiolaria striatula*. *Prymnesium parvum* is a widely distributed phytoflagellate found in several oceans, in tide pools, estuaries and in brackish waters. The first experiments on its toxicity were carried out by Liebert and Deerns in 1920, following the mass death of fishes in the Workum-See. The organism was subsequently identified by Carter (1938). *P. parvum* became of further concern in 1957, when a bloom of the flagellate threatened the commercial fish-breeding industry in Israel. Subsequently, Droop (1954) in Scotland and Reich and Kahn (1954) in Israel isolated and grew the organism in axenic cultures. In 1958, Yariv attempted to purify the toxin(s) of the organism and, subsequently, he and Hestrin (1961) suggested the name "prymnesin" to denote the extracellular toxin derived from water samples rich in *P. parvum*. The toxin was concentrated from culture filtrates and pond waters by adsorption on $Mg(OH)_2$. Further purification was obtained by treatment with acetone and methanol.

Shilo and Rosenberger (1960) extracted the toxin with methanol from dry cells concentrated by centrifugation. Paster (1968) purified a toxic prymnesin by concentrating the cells by centrifugation, removing the pigments with acetone and treating the crude prymnesin with methanol. The toxin was precipitated by ether, dissolved in water and fractionated on Sephadex G-100. Following several additional steps, Paster obtained a toxin with a molecular weight of approximately 23 000 and elementary analysis of 42·5% C, 6·95% H, 50·55% O; lipid constituents of about 30%; sugar constituents of about 70%; and intraperitoneal $LD_{50}$ for mice of 1·5 mg/kg body weight and a haemolysis index for rabbit erythrocytes of 25 mg/ml.

Ulitzer and Shilo (1970), using a different separation technique, obtained a second toxin called "toxin B" which had 15 amino acids and a number of unidentified fatty acids. In contrast to prymnesin, toxin B resembled a proteolipid. It had six haemolytic factors. Padilla and Martin (1973) summarized the preparation of *P. parvum* toxin. The toxin is isolated from the organism grown in artificial seawater media (10%) maintained at 25°C under constant illumination. The sea water is enriched with liver fusion, glycerol and vitamins $B_1$ and $B_{12}$. After 5–7 days growth of the cells, they are collected by centrifugation, ground and the pigment removed with acetone. The acetone and insoluble residue are extracted with methanol and placed on a Sephadex LH-20 column. The fractions are then collected and tested for haemolytic activity. More recently Kim and Padilla (1977) have applied their separation technique for *G. breve* to *P. parvum* (see Fig. 1). Using this procedure they obtained six fractions, four of which were haemolytic and one of these was toxic to fish, but only in the presence of spermine or at pH's lower than 9.

The ability of the toxin to haemolyze rabbit red cells is the most often used assay for measuring the toxicity of prymnesin, as well as some other protistan toxins. The assay is said to be 100 times more sensitive than the ichthyotoxic test (Paster, 1973), although it is not quite clear what mechanism is involved in the ichthyotoxic test. The haemolytic activity was said to be associated with membrane interaction leading to lysis, causing loss of ions, amino acids and macromolecules (Dafni and Gilberman, 1972). The mechanism of the lytic action of *P. parvum* toxin has been studied by Imai and Inoue (1974). Using liposomes as a model membrane system, they showed severe damage to liposomes containing cholesterol but none to those without cholesterol, which is rather different from the findings of Ulitzer and Shilo (1970) who found that *Prymnesium* toxin lysed spheroplasts or protoplasts of bacteria whose membranes lacked cholesterol. The problem is unresolved.

The crude toxin evokes a slow contraction of the guinea-pig ileum, which is caused by the release of acetylcholine, and block of the contractions induced by smooth muscle stimulants (Bergmann *et al.*, 1964). It is said to have no effect on the rat uterus. The mechanism(s) involved in the smooth muscle changes have not been established. When studied on the frog heart, the toxin causes a shortened action potential followed by a block in diastole, with marked depolarization. It is generally thought that the effects on the frog heart and guinea-pig ileum are caused by the same factor.

In the frog sartorius nerve-muscle preparation, there is a decrease in the indirectly elicited contractions on addition of the toxin to the bath. The directly elicited contractions are little affected and the nerve still propagates action potentials. End plate potentials disappear with time. It has been suggested that the toxin acts as a non-depolarizing blocking agent on the postsynaptic membrane of the endplate. In the deep extensor abdominal medialis muscle (DEAM) preparation of the crayfish, a non-cholinergic junction, the block was also found to be at the neuromuscular junction (Parnas and Abbott, 1965).

It is well known that in addition to the effects described above, the toxin alters a variety of mammalian and non-mammalian cells in cultures, bacteria, Ehrlich ascites cells and HeLa cells. It seems quite possible that the basic mechanism(s) involved in these tissue and cell changes is the same as that of those involved in the haemolysis; that the cytolytic, haemolytic, ichthyotoxic, neurotoxic, cardiotoxic, etc. changes may be due to a common mechanism.

The marine benthic green algae (Chlorophyta) are the organisms usually responsible for green tides or blooms. Several species are of considerable importance to man, since they provide part of his diet, particularly in the Phillipines and other parts of the Orient. A usually edible benthic genus from the Phillipines, *Caulerpa* sp., is known to be toxic during the rainy months, and injury to the plant thallus causes extrusion of toxin. The two toxins from *C. racemosa* and certain other species are known as caulerpicin and caulerpin (Aguilar-Santos and Doty, 1968; Maiti *et al.*, 1978).

$$CH_3(CH_2)_{13}-\underset{\underset{HN-CO-(CH_2)_n-CH_3}{|}}{CH}-CH_2OH$$

$$n = 23, 24, 25$$

Caulerpicin

Caulerpin

These toxins are transferred through the food-chain to sea snails, soft corals, or by eating the dried alga. In humans the toxins produce paraesthesias about the mouth, tongue and terminal parts of the extremities, often as a feeling of coldness. Vertigo, ataxia and respiratory distress may also occur. The clinical manifestations are self-limiting and usually disappear within 12 h.

The green alga *Cheatomorpha minima* is toxic to fishes and has haemolytic activity (Fusetani *et al.*, 1976); little is known of its definitive biological properties. Another green alga, *Ulva pertusa*, also has several haemolytic fractions (Fusetani and Hashimoto, 1976), two being water soluble and one fat soluble. The fat-soluble haemolysin is palmitic acid, a $C_{16}$ saturated fatty acid with a haemolytic activity of 0·24 saponin units/mg. One of the water-soluble haemolysins is thought to be a galactolipid, $C_{31}H_{58}O_{14}$, while the other is believed to be a sulpholipid, $C_{25}H_{47}O_{11}SK$ (Hashimoto, 1979).

"Mozuku" poisoning is caused by some brown marine algae of the families Chordariacae and Nemacystaceae. The poison may also be found in some red algae. Two fat-soluble fractions, "A and B", prepared from *Sphaerotrichia divaricata* and *Cladosiphon okanuranus* were isolated as shown in Fig. 3.

The $LD_{50}$ of the toxins was not determined but toxin A had an intra-peritoneal lethal dose of 250 mg/kg body weight. The toxins are produced

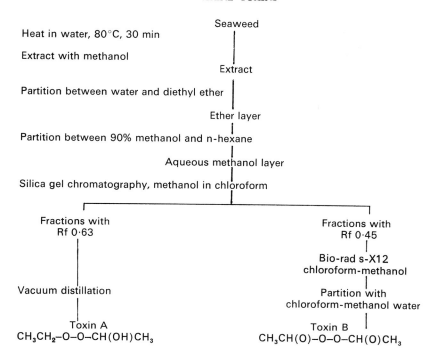

FIG. 3. Separation of fat soluble toxins from *Sphaerotrichia divaricata* as modified from Fusetani and Hashimoto (1981).

only on heating in water. Fusetani and Hashimoto (1981) suggest that the peroxides are responsible for mozuku poisoning, since the poisoning occurs after eating a hot water infused preparation.

The red algae (Rhodophyta) and the sponges contain numerous halogenated metabolites, particularly bromo compounds. The red alga *Asparagopsis toxiformis* contains many halo-compounds having 1–4 carbons. The major algal component is bromoform but the organism also contains the antiseptic, iodoform, the iodo analogue of the war gas phosgene, halogenated acetones, butenones and other substances (Burreson *et al.*, 1976). *Asparagopsis* is an edible alga and used as a seasoning for foods in part of the Pacific.

Halogenated monoterpenes and farnesylacetone epoxide have been isolated from the red alga *Plocamium* and the brown alga *Cystaphora*, respectively. In mice these fractions depress polysynaptic spinal reflexes and display weak anticonvulsant activity (Spense *et al.*, 1979). It was, therefore, surprising that these authors had found that plocamodiene A from *Plocamium cartilagineum* (L.) produced a long lasting but reversible tonic extension of the hind limbs following oral or intraperitoneal administration of the toxin to mice (Spense, 1978).

Plocamodiene A

This response was antagonized by diazepam but not the other muscle relaxants tested. It was suggested that plocamodiene A selectively affects reflexes associated with non-muscle afferent fibres (Taylor and Spense, 1979).

In many of the studies on protista, comparisons have been made on the basis of an intraperitoneal or even intravenous $LD_{100}$, the minimal lethal dose, or other unfortunately equally unreliable end points (Trevan, 1927; Russell, 1966a). In addition, some of the basic principles of pharmacology have been overlooked, resulting in misleading toxicological premises (Russell, 1980a). Signs such as a mouse dragging its hind limbs, or even its tail, or jumping during an agonal period is often tagged as the effect of a "neuro-toxin", and, as has so often occurred, another investigator finds the same substance to have an effect on red blood cell membranes, and chooses to call it a "haemotoxin," or "cytotoxin." An unnecessary and distracting dispute sometimes ensues. It is well to repeat that until we have studied the effects of a toxin on many tissues, as well as on intact preparations, great care should be taken in labelling substances as neurotoxins, haemotoxins and the like. This is not to imply, however, that a toxin cannot be said to have neurotoxic, haemotoxic or other activities but to express caution until we can differentiate between site and mechanism phenomena.

## III. Porifera

Sponges are among the simplest of animals. They are highly organized colonies of unicellular nomads composed of loosely integrated cells covered by a skin and, with few exceptions, supported internally by a skeleton of silica, calcite, or spongin. There are more than 5000 species and they are found in almost every sea from mid-tide levels to the deepest parts of the oceans. Members of the family Spongillidae are found in fresh water to 3350 m above sea level. When sponge cells are squeezed to the point where a milky fluid is obtained, examination in sea water reveals a mass of disassociated cells, moving freely at first but then collecting in groups and, eventually, coalescing into succes-sively larger groups until small but functional sponges are formed (Burton, 1961). This technique has been used in attempting to cultivate sponge toxins, some biologically active components of sponges, and to study antigen–

antibody relationships but, to date, it has not been too successful. For the most part the toxic sponge species are marine.

It has long been known that some sponges release a toxic substance into their environment. De Laubenfels (1932) observed that when *Tedania toxicalis* de Laubenfels was placed in a bucket with fishes, crabs, molluscs and worms, in an hour or perhaps less the animals would be found dead. Although this phenomenon has usually been considered as a purely defensive reaction initiated when the sponge became endangered, Green (1977) suggested that the toxic material may be released as a continuous product into the surrounding water, and thus serve as a warning or deterrent to an approaching predator. This seems a reasonable hypothesis. It may be that quantitative differences in the release of toxin are involved, perhaps initiated by physical or chemical stimuli. The idea of continuous elaboration of the toxin has elicited specific studies including those by Janice Thompson at the Scripps Institution of Oceanography who, working with *Leiosella* sp. has found that this sponge does secrete the toxin as a continuous process (unpublished results).

Although elaboration of the toxin has sometimes been noted as a "by-product of metabolism", this theory seems strained, in view of the fact that whether or not a continuous process is involved it would appear inconsistent that nature would depend on a by-product for the specificity of function the secretory material appears to serve in the animal's posture. While there seems to be little doubt that the toxin has a defensive function, Bakus (personal communication, 1982) suggests that in addition to the fish or other predator repelling function the toxin may prevent the settling of larvae or spores on the animal's surface, and it may also prevent overgrowth of one species by another. A final point that speaks against the secretion being a by-product of metabolism is the amount of energy expended in its production.

Much of the work on the ecology of toxic sponges has been carried out by Bakus and his colleagues at the University of Southern California (see Bakus, 1969). In his various works he has suggested that as the diversity and biomass of fishes increases, as in tropical waters as compared to temperate or polar waters, there is increased competition for food which, in turn, gives rise to exploitation of food sources. This leads to more specialized feeding habits. Such activity would appear to stimulate some type of natural selection in sponges, such as chemical defence through retardation or, in some cases, prevention of predation or grazing by fishes; diverse cryptofauna, animals hiding from fish predators among coral branches and the like; or changes in structure, etc. These among other mechanisms might be involved in the natural selection process by which the sponge survives.

The natural selection of chemical defences has more recently been demonstrated by Bakus (1981), who found that 73 % of all exposed common reef invertebrates on the north Great Barrier Reef were toxic to fish. Sixty per cent of the exposed sponge species were toxic, while only 33 % of the cryptic species were toxic.

It is well known that a few fishes belonging to highly specialized teleost families feed on toxic porifera and are not affected (Randall and Hartman, 1968). These species also consume large quantities of non-toxic sponges. Perhaps, their intake of toxic sponges is not sufficient to cause poisoning or the process of detoxification in these fishes is more highly developed. It is suggested that the toxic sponge species, being eaten in only small numbers or amounts are thus protected from rapid total depletion. In his work on fishes, Green (1977) demonstrated that the toxicity of sponges increased with decreasing latitudes and that the most commonly exposed sponges were the most toxic to fishes. The same is true of certain other marine invertebrates (Bakus and Green, 1974). Some nudibranchs feed exclusively on both toxic and non-toxic porifera without deleterious effects (Graham, 1955) as do, perhaps, some annelids, molluscs, bryozoans, and crustaceans. The feeding relationships between these animals and sponges is not well understood.

## A.  Sponge Poisoning

With respect to humans, poisoning probably occurs through deposit of the toxin(s) in the very superficial abrasions produced by the fine, sharp spicules of the sponge. It has been known since the days of Pliny that traumatic injury to the human skin can be produced by the spicules, particularly those of the hexactinellids, and it is believed that in many cases of poisoning this occurs prior to the deposit of the poison on the skin. Certainly, an abraded skin is more likely to absorb a toxin than an uninjured one. In 1965a I reported a case of poisoning involving a 27-year-old skin diver, who having abraded the skin of his hands while collecting stony, but not otherwise dangerous coral, decided to assist in the packing of fresh *Tedania nigrescens* (Schmidt) aboard the boat. After handling these sponges for approximately 30 min, he complained of an intense burning sensation over the hands, pruritus, and malaise. When seen 1 h later, the patient presented systemic manifestations. This case demonstrates that sponge poisoning can occur on handling the animal after it has been removed from the water (Russell, 1965a,b). Bakus (personal communication, 1982) states that the pain provoked by contact with *Neofibularia nolitangere* (Duchassaing and Michelotti) is considerably more severe than that produced by *T. nigrescens*. Das *et al.* (1971) have reported that contact with *Suberites inconstans* results in localized pruritus and minor swelling. "Sponge fisherman's disease" of the eastern Mediterranean (maladie des pêcheurs d'éponges), described by Zervos (1934), is caused by the Actiniae frequently found with sponges.

## B. Chemistry and Pharmacology

Many sponges have an offensive odour and taste but what part these qualities play in defence or in poisoning is not known. Although studies on the chemistry of sponges have progressed with great rapidity during recent years, specific chemical and pharmacological investigations on the toxic components have lagged far behind. Table III shows some of the marine sponges of medical or toxicological importance. The reader should consult Halstead (1965), Jakowska and Nigrelli (1960), Stempien et al. (1970), Green (1977), Bakus and Thun (1979) for more detailed data on the toxicity of these species. It should be noted that some sponges of the same genus as those found to be toxic to fishes have also been found to be non-toxic, and some species have also been found to be both toxic and non-toxic. The family Haliclonidae appears to have the most consistently toxic species.

TABLE III. SOME MARINE SPONGES WHICH HAVE BEEN REPORTED TO BE TOXIC TO HUMANS (A) OR WHOSE TOXIC PROPERTIES HAVE BEEN STUDIED (B)

| Name | Distribution |
| --- | --- |
| **(A)** | |
| *Dysidea etheria* De Laubenfels | West Indies, Florida |
| *Haliclona viridis* (D. & M.) | West Indies, North Carolina, Florida, ?West Central Pacific, |
| *Hemectyon ferox* (D. & M.) | West Indies |
| *Ircinia felix* (D. & M.) | North Carolina to Venezuela, ?Mediterranean |
| *Microciona prolifera* (Ellis & Solander) | Atlantic coast of North America, Willapa Bay, Washington; San Francisco Bay, California |
| *Neofibularia nolitangere* (D. & M.) | West Indies, Florida |
| *Pseudosuberites pseudos* Dickinson | Gulf California and Mexico |
| *Spheciospongia vesparium* (Lamarck) | West Indies, Florida |
| *Spinosella vaginalis* (Lamarck) | West Indies |
| *Spirastrella inconstans* (Dendy) | Indo-Pacific |
| *Suberites domunculus* (Olivi) | Mediterranean, Black Sea, Northwest Africa |
| *Tedania nigrescens* (Schmidt) | Cosmopolitan |
| *Tedania toxicalis* De Laubenfels | California |
| | |
| **(B)** | |
| *Adocia* sp. | |
| *Agelas dispar* D. & M. | West Indies, Florida |
| *Callyspongia* sp. | |
| *Callyspongia fallax* D. & M. | West Indies, Florida, ?South Australia |

TABLE III. (CONTINUED)

| Name | Distribution |
|---|---|
| *Calyx nicaeensis* (Risso) | Mediterranean |
| *Chondrilla nucula* Schmidt | West Indies, Florida, Adriatic Sea, Ligurian Sea |
| *Dysidea etheria* De Laubenfels | West Indies, Florida |
| *Geodia* sp. | |
| *Geodia cydonium* (Jameson) | Mediterranean, European seas, West Coast Africa |
| *Geodia gibberosa* Lamarck | North Carolina, West Indies, Guiana |
| *Geodia mesotriaena* Lendenfeld | Alaska to Gulf California |
| *Halichondria okadai* (Kadota) | Japan, Korea |
| *Halichondria panicea* (Pallas) | Cosmopolitan |
| *Haliclona* sp. | |
| *Haliclona doria* De Laubenfels | West Indies |
| *Haliclona erina* De Laubenfels | West Indies, Brazil |
| *Haliclona rubens* (Pallas) | Florida, West Indies, Gulf Mexico |
| *Haliclona viridis* (D. & M.) | Florida, West Indies, North Carolina, Gulf Mexico, ?West Central Pacific |
| *Hemectyon* sp. | |
| *Hymeniacidon ?amphilecta* De Laubenfels | Florida, West Indies, Gulf Mexico |
| *Hymeniacidon perlevis* (Montagu) | Mediterranean, East Atlantic, Indian Ocean, ?Australia, ?Japan |
| *Ianthella* sp. | |
| *Iotrochota birotulata* (Higgin) | Florida, West Indies, Indo-Pacific |
| *Ircinia campana* (Lamarck) | Gulf Mexico, Florida, West Indies, Brazil |
| *Ircinia felix* (D. & M.) | N. Carolina, Florida, Gulf Mexico, West Indies, Caribbean, Venezuela, ?Mediterranean |
| *Ircinia strobilina* (Lamarck) | Florida, West Indies |
| Keratose sponge | |
| *Xestospongia subtriangularis* (Duchassaing) | Florida, West Indies |
| *Lissodendoryx* aff. *kyma* De Laubenfels | California to Washington |
| *Microciona parthena* De Laubenfels | California |
| *Microciona spinosa* Wilson | Florida, West Indies |
| *Mycale* sp. | |
| *Mycale lingua* (Bowerbank) | Alaska to Washington and Newfoundland to Maine |
| *Niphates erecta* D. & M. | N. Carolina, Florida, Caribbean |
| *Pseudoceratina crassa* (Hyatt) | Florida, West Indies |
| *Ridleia* sp. | |
| *Sigmadocia* sp. | |
| *Smenospongia* sp. | |

| | |
|---|---|
| *Spongia officinalis* L. | Mediterranean |
| *Tethya actinia* De Laubenfels | West Indies, ?Marshall Islands |
| *Tethya aurantia* (Pallas) | Cosmopolitan |
| *Thalyseurypon* sp. | |

From Russell, 1965; Halstead, 1965; Hashimoto, 1979 and G. J. Bakus (personal communication, 1982).

As for so many of the marine poisons, the study of sponge toxins has lacked a standardized assay method which has applied significance. However, since toxicity was first noted as a deleterious effect to fishes exposed to sponge excretions, the primary assay has involved observations on the effects of crude sponge extracts on fishes in a tank, or merely the observation of fishes exposed to live sponges in a small aquarium.

Perhaps the most extensive studies of such observations have been those by Bakus and his colleagues. Essentially, the shipboard preliminary assay method now used by this group involves grinding 5 g of the sponge in 10 ml of sea water, centrifuging, pouring the supernatant into a bowl with 300 ml of sea water, and then placing a 1·5–5 g sergeant major (*Abudefduf saxatilis*) into the water and observing the fish's behaviour over a designated period of time. In preliminary testing in the laboratory, a similar programme is followed, except 200 ml of tap water and 1·2–2.4 g goldfish (*Carassius auratus*) are used.

After preliminary testing, a standard laboratory bioassay is used. Serial dilutions of the crude extract are prepared, starting with a 0·1 g crude material/ml tap water; an alcohol extract is used for comparison. Water temperature in the 11-litre aquarium is 22–27°C and the $LC_{50}$ (lethal concentration for 50% of the animals) determinations are done for the more highly toxic sponges, using an alcohol extract. Toxicity is determined on the basis of the fishes swallowing air, blowing bubbles, being bitten by normal fish, equilibrium loss, erratic swimming behaviour, slow swimming movements, escape responses, thrashing behaviour, extreme lethargy or stupor, failure to recover when put in fresh water, and death (Bakus and Thun, 1979).

In a study of 54 species of Caribbean sponges, it was found that 31 species were toxic to fish. The most toxic species was *Hymeniacidon ?amphilecta* de Laubenfels, with species of *Haliclona* being less toxic and *Ireinia* sp., Microcronidae, *Neofibularia nolitangere*, *Xestospongia muta* (Schmidt), also being less toxic (Bakus and Thun, 1979).

In 1906, Richet precipitated a substance from extracts of the siliceous sponge *Suberites domunculus*, which when injected into the dog produced vomiting, diarrhoea and dyspnea, and caused haemorrhages in the gastric and intestinal mucosa, peritoneum and endocardium. The lethal dose in dogs was 10 mg/kg and the toxic substance was found to be non-toxic when administered orally. The poison was called "suberitine" (Richet, 1906;

Lassabliere, 1906). Arndt (1928) demonstrated that extracts from certain freshwater sponges produced diarrhoea, dyspnea, prostration and death when injected into homoiothermic animals. These same extracts had some haemolytic effect on sheep and pig erythrocytes, and blocked cardiac function in the isolated frog heart preparation. The extracts were heat stable and produced no deleterious effects when taken orally.

Das *et al.* (1971) found that extracts of *Suberites inconstans* Dendy produced a histamine-like effect on the guinea-pig intestine and attributed this to histamine, which they found in the sponge. On paper chromatography they detected five other amines, three having phenolic groups. Dried specimens of *Fasciospongia cavernosa* yielded crystals of *N*-acyl-2-methylene-β-alanine methyl esters. In mice, the subcutaneous lethal dose of the crystals was approximately 120 mg/kg body weight (Kashman *et al.*, 1973), obviously not very toxic by toxicological standards. Algelasine from *Agelas dispar* has some activities of a saponin. A unique sesquiterpene, 9-isocyanopupukeanane, has been isolated from the nudibranch *Phyllidia varicosa* and has been found to be present in the sponge *Hymeniacidon* sp., on which the nudibranch feeds (Burreson *et al.*, 1975).

Sesquiterpene

The sesquiterpenes, and some isonitriles and isothiocyanate sesquiterpenes are known to be toxic to various animals on injection. The reader is referred to Hashimoto's (1979) text for a more thorough review of the properties of these substances.

Cariello *et al.* (1980) isolated and characterized Richet's suberitine. Living specimens were squeezed, the extract centrifuged in the cold at 27 000 × g for 1 h, the sediment resuspended in sea water and centrifuged again, the two supernatants exposed to cold ethanol, and the precipitate removed by centrifugation. The sediment was then extracted with sodium acetate, the suspension centrifuged, and the precipitate re-extracted with the same buffer. The procedure was repeated until the supernatant displayed no lethal activity; concentration was by ultrafiltration. The product was next fractionated on Sephadex G-150 and the active fraction, which caused paralysis when injected into the crab, was chromatographed on Sephadex G-75 and further purified on Sephadex C-50. Homogeneity was demonstrated on SDS polyacrylamide gel electrophoresis, and the molecular weight (28 000) was estimated by ultracentrifugation. The toxin had a marked haemolytic effect on human erythrocytes and some ATPase activity. Studies on the giant axon of the abdominal nerve of the crayfish showed that in a concentration 4·4 mg /ml there was depolarization, followed by an irreversible block in the indirectly stimulated action potential. The authors speculated that this irreversible block may explain the flaccid paralysis seen in crabs following injection of suberitine into the arthropod's haemolymph.

Wang *et al.* (1973) demonstrated that a preparation of an extract of *Haliclona rubens* exerted a depolarizing action on the end-plate membrane of the frog skeletal muscle and that a lesser depolarization occurred in the membrane elsewhere than at the end plate. This activity differed from that

TABLE IV. SUBSTANCES ISOLATED FROM SPONGES

| | |
|---|---|
| Amino acids | Nucleosides |
| Lysine | Ribonucleic acid |
| Betaine | Desoxyribonucleic acid |
| Taurine | -β-D-arabofuranoside of thymine |
| Hypotaurine | -β-D-arabofuranoside of uracil |
| Taurobetaine | -β-D-ribofuranoside of 2-methoxyadenine |
| Taurocyamine | Pentofuranoside of uracil |
| Histamine | Guanine |
| Dimethylhistamine | Adenine |
| Agmatine | *O*-methyl purine |
| Guanidine derivatives | Methyladenine |
| Glycocyamine | 1-methyladenine |
| Putrescine | Spongouridine |
| Phosphocreatine | Spongothymidine |
| Phosphoarginine | Choline |
| Hippospongine | Acetylcholine |
| Zooanemonine | Cholesterol |
| Herbipolin | Cholestanol |
| Eledonine | Neospongosterol |
| Halitoxin | Clinasterol |
| Biogenic amines | Poriferasterol |
| Halogenated dibromotyrosine-derived | |
| compounds | Homarine |
| Collagens | Ubiquinone |
| Glycoproteins | Isoprenoids |
| Pentosides | Squalene |
| Acid polysaccharides | Lipids |
| Inosite | Sterols |
| Inositol | Terpenoids |
| Prenylated benzoquinones | Sesquiterpenoids |
| Bromopyrrole derivatives | Carotenoids |
| Agelasine | Manoalide |

caused by batrachotoxin and grayanotoxin (a toxin from the plant Ereaceae). The crystalline toxin from *Halichondria okadai*, okadaic acid, has been isolated and characterized. It is a cytotoxic monocarboxylic acid with the molecular formula of $C_{44}H_{68}O_{13}$. Its intraperitoneal $LD_{50}$ in mice is 0·19 mg/kg body weight. Spectroscopic studies indicate it is an ionophoric polyether and diffraction crystallography of the *o*-bromobenzyl ester indicates a heptacyclic structure (Tachibana, 1980).

It seems advisable to include, in any review of sponge toxicity, a short discussion on the general chemistry and pharmacology of sponges, since, although most of the substances isolated from sponges have not been studied for their specific pharmacological activities there is the possibility that some of them may have important toxicological properties. In addition, the sponges are the first marine invertebrates to be studied for the presence of biologically active compounds in a systematic and definitive way, and from the chemical standpoint they have certainly been the most intensively studied of all the marine invertebrates. Table IV shows some of the substances that have been identified in marine sponges. No attempt has been made to categorize these or place them in any specific order, since there is considerable overlapping and some duplication, some terms are now obsolete, and the specific chemical nature of most is not clearly enough established to classify them.

The interest in the chemical structure of sponges arises from several factors: (1) more than any other marine animal they are the source of novel chemical compounds; (2) their chemical constituents, to date, appear to have the greatest potential as tools in the study of biological mechanisms; (3) they are a rich source of primitive organic compounds and secondary metabolites; and (4) they have been shown to have potentially useful bacteriostatic, antibiotic, antifungal, antiviral, antimitotic, anti-inflammatory and, possibly, antitumoural properties.

The current period of sponge chemistry and pharmacology started with the isolation of two nucleosides, spongouridine and spongothothymidine from the Jamaican sponge *Tethya crypta* (Bergmann and Feeney, 1951).

Spongouridine                    Spongothymidine

These nucleosides occur free in sponges and have served as models for synthesizing the chemical analogue D-arabinosylcytosine, a nucleoside having antiviral and certain other pharmacological activities that have been attributed to its property of inhibiting pathways in nucleic acid biosynthesis (Cohen, 1963). Free amino acids have been identified in abundance in sponges, including some relatively rare naturally occurring ones, such as β-amino isobutyric and pipecolic acids (Bergmann, 1962; Berquist and Hogg, 1969; Berquist and Hartman, 1969). The free amino acids have been of particular importance in establishing fingerprints for taxonomic problems within the Porifera.

Although much work has been done on the sterols of sponges, particularly with respect to structure-systematics relationships, definitive toxicological studies are lacking. Poriferan sterol chemistry owes its beginning to the work of Bergmann (1949), and with the advent of thin-layer and gas chromatography, mass spectrometry and other isolation and characterization techniques it has advanced remarkably during the past decade. Berquist (1978) states that the modern period of reinvestigation of sponge sterols dates from 1972. She notes among the novel sterol types the astysterols, with new side chain alkylation patterns; the calysterols, having a cyclopropene ring in the side chain; and the norsteranes, with a modified tetracyclic ring structure. The sponges which have a moderate to high proportion of lipids in the form of sterols contain relatively small amounts of terpenes, and the reverse condition frequently applies. Although the sterols of sponges may not be particularly toxic, if at all, some investigators have suggested that they may play a part in the metabolism of the toxic component(s).

Berquist (1978) also notes that while sponges contain numerous biologically active substances, four groups are of particular interest: the terpenoids, heavily halogenated dibromotyrosine-derived compounds, bromopyrrole derivatives, and the prenylated benzoquinones. The terpenoids are, for the most part, linear furans, such as the antibiotic "furospongin-1" and the sesterterpene, "furospongin-3" from *Spongia officinalis*.

Furospongin-1

Furospongin-3

The terpenoids are highly astringent to the taste and may cause the sponge to be unpalatable to predators.

The dibromotyrosine-derived compounds are small molecules containing bromine. *Verongia* sp. are of particular interest because of their potent antibiotic activity (Cimino *et al.*, 1975). Their toxicological properties have not been studied. The bromopyrrole compounds have been found in at least five species of three genera, *Agelas*, *Axinella* and *Phakellia* (Cimino *et al.*, 1975). "Oroidin", obtained from *Agelas oroides* and certain other species is a more complex bromopyrrole metabolite. Its toxicological properties are not known. The prenylated benzoquinones of sponges are a novel group whose pharmacological properties are of interest because related compounds in other marine invertebrates function to confuse the olfactory sense of predators (Kittredge *et al.*, 1974). Although such a function has not been demonstrated for the sponge quinones, Berquist (1978) believes such research might prove rewarding.

# IV. Cnidaria

The phylum Cnidaria or Coelenterata (hydroids, jellyfish, sea anemones and corals) are simple metazoans that possess the two basic tissues found in all higher animals, a layer of jelly-like material with supporting elastic fibres between the ectoderm and endoderm known as "mesogloea", a gastrovascular cavity that opens only through its mouth, radial symmetry and tentacles bearing abundant nematocysts. In the Portuguese man-of-war, *Physalia*, and in many other cnidarians the tentacles contain long muscle strands which can be contracted to bring the animal's prey to the feeding polyps below the umbrella. These polyps engulf the prey and digest it. Venomous forms are found in all three classes of living cnidarians: Hydrozoa, or hydroids, hydromedusae and fire corals; Scyphozoa, or true jellyfish; Anthozoa, or sea anemones, sea feathers and corals.

The Hydrozoa are branched or simple polyps, some having budded medusae, usually found growing as tufts on rocks, pilings or on some seaweeds. The order Siphonophora includes the Portuguese man-of-war, *Physalia*. Most of the hydroids are marine. The Scyphozoa, true medusae or jellyfish, are typified by a body, umbrella, or bell, which is usually convex above and concave below. They lack a velum and stomodaeum, and the polyp is either reduced or absent. The Cubomedusae or sea wasps are the most dangerous of all the cnidarians, particularly *Chironex fleckeri* Southcott and *Chiropsalmus quadrigatus* Haeckel of Australia. All species are marine. Finally, the class Anthozoa contains the corals, sea anemones and alcyonarians. The corals have calcareous skeletons and are our reef builders. The anemones are sedentary flower-like structures. The alcyonarians include the stony, soft,

TABLE V. SOME VENOMOUS CNIDARIANS

| Class | Family | Genus and species |
|---|---|---|
| **HYDROZOA** | | |
| Medusae | Corynidae | *Sarsia tubulosa* (Sars) |
| | Pandeidae | *Leuckartiara gardineri* Browne |
| | Geryonidae | *Liriope tetraphylla* (Chamisso & Eysenhardt) |
| | Olindiadidae | *Gonionemus vertens* (Agassiz) |
| | | *Olindias sambaquiensis* Müller |
| | | *Olindias singularis* Browne |
| | | *Olindioides formosa* Goto |
| | Pennariidae | *Pennaria tiarella* (Ayres) |
| Hydroids | Halecidae | *Halecium beani* (Johnston) |
| | Plumulariidae | *Aglaophenia cupressina* Lamouroux |
| | | *Lytocarpus pennarius* (Linnaeus) |
| | | *Lytocarpus philippinus* (Kirchenpauer) |
| | | *Lytocarpus phoeniceus* (Busk) |
| Millepore corals | Milleporidae | *Millepora alcicornis* Linnaeus |
| | | *Millepora complanta* Lamarck |
| | | *Millepora dichotoma* Forskål |
| | | *Millepora platyphylla* Hemprich & Ehrenberg |
| | | *Millepora tenera* |
| Siphonophores | Physaliidae | *Physalia physalis* Linnaeus |
| | | *Physalia utriculus* (La Martiniere) |
| | Rhizophysidae | *Rhizophysa eysenhardti* Gegenbaur |
| | | *Rhizophysa filiformia* (Forskål) |
| | | |
| | | |
| **SCYPHOZOA** | | |
| Jellyfishes | Carybdeidae | *Carybdea alata* Reynaud |
| | | *Carybdea marsupialis* (Linnaeus) |
| | | *Carybdea rastoni* Haacke |
| | | *Tamoya gargantua* Haeckel |
| | | *Tamoya haplonema* Müller |
| | Cassiopeidae | *Cassiopea xamachana* R. P. Bigelow |
| | Catostylidae | *Acromitoides purpurus* (Mayer) |
| | | *Catostylus mosaicus* (Quoy & Gaimard) |
| | Chirodropidae (Cubomedusae) | *Chirodropus gorilla* Haeckel |
| | | *Chironex fleckeri* Southcott |
| | | *Chiropsalmus buitendijki* Horst |
| | | *Chiropsalmus quadrigatus* Haeckel |
| | | *Chiropsalmus quadrumanus* (Müller) |
| | | *Corukia barnesi* |

TABLE V. (CONTINUED)

| Class | Family | Genus and species |
|-------|--------|-------------------|
| | Cyaneidae | *Cyanea capillata* (Linnaeus) |
| | | *Cyanea ferruginea* Eschscholtz |
| | | *Cyanea lamarcki* Péron & Lesueur |
| | | *Cyanea nozaki* Kishinouye |
| | | *Cyanea purpurea* Kishinouye |
| | Lobonematidae | *Lobonema mayeri* Light |
| | | *Lobonema smithi* Mayer |
| | Pelagiidae | *Chrysaora helvola* Brandt |
| | | *Chrysaora hysoscella* (Linnaeus) |
| | | *Chrysaora melanaster* Brandt |
| | | *Chrysaora quinquecirrha* (Desor) |
| | | *Pelagia colorata* Russell |
| | | *Pelagia noctiluca* (Forskål) |
| | | *Sanderia malayensis* Gotte |
| | Rhizostomatidae | *Rhizostoma pulmo* (Macri) |
| | Ulmaridae | *Aurelia aurita* (Linnaeus) |
| ANTHOZOA | | |
| Corals | Acroporidae | *Acropora palmata* (Lamarck) |
| | | *Astreopora* sp. |
| | Poritidae | *Goniopora* sp. |
| Anemones | Actiniidae | *Actinia equina* Linnaeus |
| | | *Anemonia sulcata* (Pennant) |
| | | *Anthopleura xanthogrammica* (Brandt) |
| | | *Bunodactis elegantissima* (Brandt) |
| | | *Condylactis gigantea* (Weinland) |
| | | *Physobrachia douglasi* Kent |
| | Actinodendronidae | *Actinodendron plumosum* Haddon |
| | Actinodiscidae | *Rhodactis howesi* Saville Kent (poisonous) |
| | Agariciidae | *Pavona obtusata* Crossland |
| | Aiptaisiidae | *Aiptasia pallida* (Verrill) |
| | Aliciidae | *Alicia costae* (Panceri) |
| | | *Lebrunia danae* (Duchassaing & Michelotti) |
| | Alcyonidae | *Sinularia abrupta* Tixier-Durivault |
| | | *Sarcophyton glaucum* (Quoy & Gaimard) |
| | Diadumenida | *Diadumene cincta* Stephenson |
| | Hormathiidae | *Adamsia palliata* (Bohadsch) |
| | | *Calliactis parasitica* (Couch) |
| | Isopheliidae | *Telmatactis vermiformis* (Haddon) |
| | Sagartiidae | *Corynactis australis* Haddon & Duerden |
| | | *Sagartia elegans* (Dalyell) |
| | | *Sagartia longa* (Verrill) |

| Stoichactiidae | *Radianthus paumotensis* (Dana) |
| Zoanthidae | *Palythoa tuberculosa* (Esper) |
| | *Palythoa toxica* Walsh & Bowers |
| | *Palythoa caribbea* Duchassaing |
| | *Palythoa mammilosa* Lamouroux |
| | *Palythoa vestitus* (Verrill) |

horny and black corals, as well as colonial sea pens and sea pansies, all of which lack a medusa stage. All known Anthozoa are marine creatures. Table V shows some of the cnidarians of particular importance because of their stingings on man or their unusual toxicological properties.

It is well known that some marine animals, protozoa, sponges, copepods, crabs and nudibranchs can invade the tentacles of sea anemones without being stung. Symbiosis between these animals and cnidarians has been attributed to a general inhibitory effect in the anemone's nervous system, to some secretion of the anemone or to some other factor but Lubbock (1980), working with the clown fish, *Amphiprion clarkii*, and the anemone *Stechodactyla haddoni*, found that the fish achieves this protection by means of an external mucous layer, that is, a layer 3–4 times thicker than that found in other fishes. This layer is chiefly glycoprotein in nature, containing neutral polysaccharides. Lubbock has shown that the layer does not contain specific nematocyst inhibitors or excitatory substances but that it is relatively inert, causing it to be quite different from the stimulatory nature of the mucus of other fishes.

The cnidarians are of particular importance to man because of the stinging qualities of their nematocysts, which limit his underseas explorations, including scientific and recreational pursuits. I might be venturist to say that all 9000 species of cnidarians have nematocysts. Most biologists appear to prefer the more cautious statement of Robson (1972) that "nematocysts are characteristic of the phylum Cnidaria and it is difficult to think of a species where they are not essential to the animals' habit of life". Nematocysts have been classified on the basis of their structure, function and taxonomy but until Weill (1934) proposed his elaborate nomenclature for these structures there was little common agreement on forms. Weill described 17 categories of nematocysts and while these have been modified with the passing of time, in general, his approach still offers a basis for common communication. The interested reader is referred to the works of Hyman (1940), F. S. Russell (1953), Hand (1961) and Werner (1965) for a more detailed discussion of nematocyst forms.

From the functional standpoint, nematocysts have been classified as *volent*, where the tube end is closed; *penetrant*, where the tube end is open; and *glutinant*, where the tube end is open and sticky. The volent type is unarmed; its threads, when discharged, wrap around and entangle the offending animal. The penetrant type is armed with spiralling rows of spines which serve in anchoring the thread to the animal. The stylus of the thread is capable of penetrating some epithelial tissues, and venom may be discharged through its open end into the wound. The piercing ability of some penetrant type nematocysts is sufficient to puncture the chitinous cuticle of several marine animals. The glutinant type of nematocyst may respond to mechanical stimuli and can be used by the cnidarian for anchoring its tentacles during locomotion.

## A.  *Venom Apparatus*

As previously noted, the stinging unit of cnidarians is the *nematocyst*, which is a capsulated, ovoid cell varying in size from 4 to 225 μm, and containing an operculum, a long coiled tube or hollow thread, matrix and venom. The nematocyst is formed as a "metaplasmic organelle" within an interstitial cell, the *cnidoblast*. These cnidoblasts are distributed throughout the epidermis, except on the basal disc. They are particularly abundant on the tentacles and are used as both offensive and defensive weapons, as well as for anchorage. The cnidoblasts are produced at a distance from their final resting site in the epithelium; none originate in the tentacles. They migrate to their final location in the ectoderm by amoeboid activity and passive transport. The cell adjusts itself to a superficial position with that part containing the nematocyst directed so the thread can be discharged into stimulating organisms.

Extending from the free surface of the apex of the nematocyst is the *cnidocil*, or cnidocil apparatus. This is a cilium-like structure, varying in length in different species but being approximately 2·5 μm in *Physalia physalis*. It is composed of closely packed microtubules, a basal plate and a striated rootlet. The cnidocil apparatus is thought to be the sole receptor and transducer for the discharge of the nematocyst. The cnidocil in hydrozoans, the ciliary-cone complex in anthozoans and the flagellum-stereo-ciliary complex in scyphozoans are thought to be homologous sensory receptors for nematocyst discharge (Cormier and Hessinger, 1980a).

The question of whether or not a single or multiple mechanism is involved in the transmission of a stimulus from the cnidocil to the cell in all species of cnidarians has not been fully answered, in spite of many excellent studies on this problem. The initiating stimulus can be either mechanical or chemical. In the laboratory it can be electrical. But whether or not the cnidocil transmits the message to the fibrillar collar, which exerts pressure on the capsule, deforming it and causing the operculum to be dislodged, as suggested by Cormier and Hessinger for *P. physalis*; or whether or not the cnidocil transmits the stimulus to a neurite or neurosecretory cell at the base of the cnidoblast, as first described by Lentz and Barrnett (1965) for *Hydra*, is not easily answered. Indeed, the matter of nematocyst discharge may be due to several factors, as well as being more complex than generally believed, and it probably varies with different species of cnidarians. The interested reader is referred to the fine reviews of Robson (1972) and Yanagita (1973) for a more definitive review of this problem.

The coiled tubule in the undischarged nematocyst varies in length from 50 µm to over 1 mm, depending on the species of cnidarian. When discharged the operculum is released and the everted tubule explodes, remaining attached at the original site of the operculum. The thread is continuous with the capsule wall, while the capsule itself is secured to a fibrillar network of microfilaments and microtubules that extend from the apex of the cell to the basal attachment of the acellular mesogloea. The structure of the tubule varies with different species. In *P. physalis* it has three spirally arranged rows of barbs that curve toward its base (Fig. 4). The open end becomes attached in the prey or victim and the venom is ejected. The manner in which the venom is ejected is not known.

The nature of nematocyst discharge and the localized fashion in which these cells respond to stimuli, whether chemical, mechanical or electrical have been the object of extensive study (Parker and Van Alstyne, 1932; Weill, 1934; Pantin, 1942; Yanagita, 1959a,b; Mackie, 1960; Lentz, 1966; Blanquet, 1972; Robson, 1972; Yanagita, 1973; Mariscal, 1974; Lubbock, 1979; and Cormier and Hessinger, 1980b), and it is not within the realm of this presentation to discuss the various theories for cell discharge. However, there is evidence to indicate that some variations exist in the mechanisms leading up to discharge and that these, for the most part, may be related to the kind of cnidarian involved. Contraction of the capsule, osmotic changes within the capsule, swelling of the capsular contents, dilatation of the thread, and several other possible mechanisms have been suggested as events involved in phenomena leading up to displacement of the operculum and explosion of the thread.

At present, studies would seem to indicate that the more probable mechanisms involve (1) changes in osmotic pressure within the capsule (the osmotic hypothesis), (2) the constant pressure hypothesis and (3) the contractile hypothesis. Cormier and Hessinger (1980b) believe that in *Physalia* the osmotic hypothesis is not applicable and suggest that the stimulus is received by the cnidocil apparatus which transmits it to the fibrillar collar with its muscle-like network of connecting microfilaments. The collar, in turn, contracts to deform the capsule, causing the operculum to be dislodged. The tubule is then discharged by the force of a spring mechanism and by hydraulic pressure. The discharge thread penetrates the prey or victim and the hooked barbs secure it, decreasing the chance of escape. As more tentacles

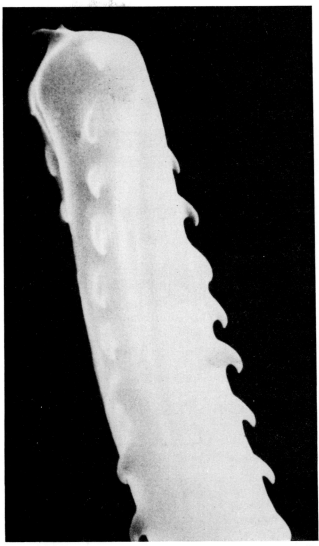

FIG. 4. Discharged everted tubule from *Physalia* showing the three rows of basally
pointed barbs (Cormier and Hessinger, 1980b).

come in contact with the prey, more nematocysts are discharged. Figure 5
shows a tubule within a nematocyst from *P. physalis*, which Cormier and
Hessinger (1980b) believe contains the venom within its canal. Figure 6
shows a nematocyst with its cnidocil, capsule contents and basal processes.

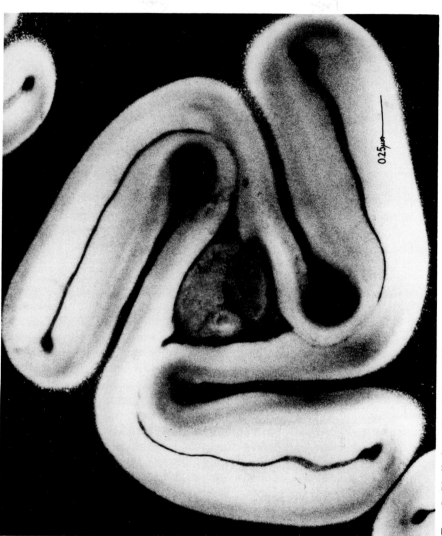

Fig. 5. Undischarged tubule within nematocyst showing canal (Cormier and Hessinger, 1980b).

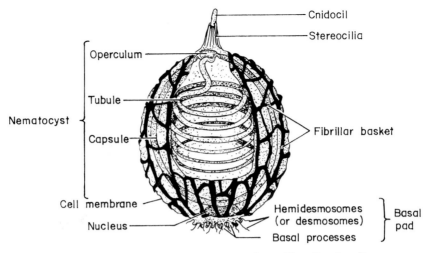

FIG. 6. Diagrammatic drawing of a nematocyst from *Physalia physalis*.

## B.  *Chemistry and Toxicology*

As in many of the earlier studies on marine venoms the chemical and toxic-ological properties of the cnidarian toxins were carried out with crude saline or water extracts prepared from the whole animal or from one or several of its parts. The findings from these studies varied considerably and the reports in the literature of those years reflect the uncertainty of the nature of cnidarian toxins. It is apparent that some of these early workers were studying normal constituents of the animal's tissues, several of which appear to be limited to tissues in the lower phyla. When these substances were injected into higher animals they produced deleterious reactions. These reactions then became aligned with clinical findings and, unfortunately, led to mis-understandings and questionable therapeutic advice (Russell, 1965a). As noted by Lane (1960), such substances as "thalassin", "congestin" and the "Cyanea principle" were probably derived from tentacular tissues rather than venom-bearing nematocysts.

With the advent of nematocyst isolation it became apparent that the chemistry of the nematocyst and that of other cnidarian body tissues were distinctly different and the original concepts about the nature of the venom as an amine or amine-like substance had to be discarded. With the works of Phillips (1956), Lenhoff *et al.* (1957), Lane and Dodge (1958), and Barnes (1967a), among others, the protein or peptide nature of coelenterate toxins became established.

Although the venoms of all cnidarians so far studied are either poly-peptides or proteins, their pharmacology has become confused, particularly with respect to the physiopharmacological relationships between the various fractions isolated by different investigators. For instance, it is not clear whether or not some of the differences in the biological activities of the various toxins are a result of the chemical techniques used in obtaining them or are due to differences in the preparation used to define the pharmacological property. Some confusion is certainly associated with the gimmick names frequently used for the various toxins, or even for the same toxin. In addition, one wonders if the proteinase inhibitor of a toxin has its sole significance, physiologically or pharmacologically, in this activity or is this property a subunit of a more applicable property in the venom and one for which our techniques are not yet geared or are not functionally applicable. These matters are not easily resolved.

The modern period of fire coral toxicology began with the works of Wittle *et al.* (1971) and Middlebrook *et al.* (1971). The former investigators studied nematocyst toxin from *Millepora alcicornis*, which they obtained by exposing the coral to a pH 5·8–7·0 phosphate buffer, precipitating with 80 % ammonium sulphate, and then passing the product through DEAE-cellulose, Sephadex G-100 and hydroxyapatite. The partially purified toxin gave three protein bands on acrylamide disc electrophoresis. From Sephadex G-100 studies a molecular weight of approximately 100 000 was suggested. The toxin was relatively stable and its intravenous $LD_{50}$ in mice was found to be 0·04 mg/kg body weight. Subsequently, using undischarged nematocysts from *M. tenera*, this group found that the toxin obtained from the undischarged nematocysts had a similar $LD_{50}$ and concluded that the toxin from the undischarged nematocysts was the same as that obtained from nematocyst discharged by exposure to the phosphate buffer. The toxin had haemolytic and dermonecrotic activities, and was antigenic with cross protection against *M. alcicornis* toxin (Wittle and Wheeler, 1974).

Middlebrook *et al.* (1971) found that they could obtain an electrophoretic-ally pure toxin from the fire coral *Millepora dichotoma* by a simple pass through DEAE-cellulose following the same initial extraction procedures. The $LD_{50}$ was 0·038 mg/kg body weight, quite similar to that found for *M. alcicornis*. The signs in mice were also similar.

The sea whip *Lophogorgia rigida* has yielded an interesting toxin known as lophotoxin. Extracts of *L. rigida* were chromatographed over magnesium silicate adsorbent and eluted with ethyl acetate in isooctane. Further puri-fication was carried out by high performance liquid chromatography. Chemical procedures gave the formula $C_{22}H_{24}O_8$. In addition, a small amount of cembrenolide, pukalide (2) was obtained. This had previously been found by Scheuer's group (1975) in the soft coral *Sinularia abrupta*.

The toxin had a subcutaneous $LD_{50}$ in mice of 8·9 mg/kg body weight and was found to block the indirectly elicited contractions in a mammalian nerve-muscle preparation, while not affecting the directly elicited contractions. In the frog rectus abdominis preparation, ACh-induced contractions were blocked but KCl-induced ones were unaffected. The authors concluded that while the studies show that lophotoxin produces an irreversible postsynaptic block, they could not exclude the possibility of a presynaptic function (Culver and Jacobs, 1981).

While techniques for separating nematocysts are not new (Glaser and Sparrow, 1909), the initial studies on a nematocyst preparation from *Physalia physalis* were performed by Lane and Dodge (1958), and by Lane (1960). Nematocysts were isolated from tentacles by autolysis at 4°C for 24–48 h and by then passing a sea-water solution of the digested tentacles through a graded screen to isolate the undischarged nematocysts, which were washed repeatedly with sea water and allowed to settle overnight. The supernatant was decanted and discarded and the residue, composed chiefly of nematocysts, was centrifuged at 300–400 rpm for 15–30 min and then resuspended in sea water. The suspension was thought to be almost completely free of tentacular tissue and contained approximately 55 million undischarged nematocysts/g wet weight (from 3·8 l of tentacles). The contents of the nemato-cysts were freed by homogenization, the homogenate centrifuged to separate the capsules and capsular fragments from the capsular contents, and the supernatant used for subsequent chemical and pharmacological testing. Using this technique the authors found the supernatant to be a highly labile protein complex, rich in glutamic acid and having an approximate intra-peritoneal lethal dose in mice of 0·037 ml/kg body weight of a preparation containing 0·02 % total nitrogen.

The toxin produced paralysis in fish, frogs and mice. Animals killed following stingings by *Physalia* exhibited marked pulmonary oedema, right cardiac dilatation with venous congestion of the larger vessels of the chest and portal circulations. It was suggested that the toxin affected the respiratory centres before producing changes in the voluntary muscles and that it altered the permeability of the capillary wall but did not produce haemolysis, although there is clinical evidence to the contrary (Guess, personal communication, 1982). It also caused changes in the isolated heart of the clam, which the authors thought resembled those provoked by acetylcholine.

In 1961, Lane subjected lyophilized "crude" extracts of *Physalia* nemato-cysts to one-dimensional chromatography and obtained nine spots, four of which accounted for 95 % of the total lethal activity in the crab *Uca pugilator*. By paper electrophoresis he separated the same extracts into four fractions, three of which contained the total lethality, the principal lethal portion being in two fractions. The crude toxin was lethal to mice at 1·7 mg/kg body weight. Lane suggested that *Physalia* toxin is a relatively simple material consisting of only a few toxic peptides which are synthesized by gastrodermal cells and which pass through the mesogloea and then into the nematocyst during the morphogenesis of this structure.

In a further study with paper electrophoresis, *Physalia* toxin was separated into four fractions, two of which accounted for 95% of the toxicity (Hines and Lane, 1962), and on polyacrylamide gel electrophoresis it was resolved into 8–10 fractions. The crude preparation had protease activity. In mice the $LD_{50}$ varied between 0·050 and 0·070 µg/kg body weight (Lane, 1967). Phospholipases A and B were detected in the crude venom but were not thought to be related to lethality (Stillway and Lane, 1971).

Various studies on crabs, rats, dogs, and a nerve-muscle preparation of the frog indicated that the toxin produced changes in the Na-K pump, resulting in depolarization of the cell membranes (Lane, 1967; Larsen and Lane, 1970). It produced a marked change of the normal pattern of pressure variations in the haemocoel and electrocardiogram of the land crab *Cardisoma guanhumi* (see Lane and Larsen, 1965). In the rat, where the $LD_{50}$ was approximately 100 µg/kg body weight, low doses of the toxin caused an increase in the Q–T interval, a decrease in the P–R interval and P wave inversion of the electrocardiogram. Large doses produced marked ECG changes leading to cardiac failure (Larsen and Lane, 1966). These authors postulated that if *Physalia* toxin causes a general sustained depolarization of the postsynaptic membrane, this might account for the differences in the responses of the mammalian (myogenic) and crustacean (neurogenic) hearts to the toxin. Subsequent studies showed that the ability of skeletal muscle sarcoplasmic reticulum to bind ionic calcium (Calton *et al.*, 1973) and that nuclear alterations and dissolution of intercellular collagen are an important cytological change in cultured hamster ovary K-1 cells (Neéman *et al.*, 1981).

Tamkun and Hessinger (1981) obtained a haemolytic protein from *P. physalis*, using nematocyst isolation as suggested by Lane and Dodge (1958). This protein, physalitoxin, was also lethal to mice at the 0·20 mg protein/kg body weight level, while the $LD_{50}$ for the crude venom was 0·14 mg/kg. A molecular weight of 212 000 was calculated. The sedimentation coefficient was 7·85 and the structure was rod-like in shape with a calculated axial ratio of approximately 1 : 10. The authors suggested that the toxin was composed of three subunits of unequal size, each of which is glycoylated. Physalitoxin is about 28% of the total nematocyst venom protein. Its carbohydrate content is 10·6% and represents the major glycoprotein of the crude venom. This haemolytic and lethal toxin was inactivated by concanavalin A. The inactivation was blocked in the presence of α-methyl-mannoside. This would seem to indicate that the inactivation by concanavalin A is probably caused by an interaction with specific saccharides on the haemolysin. The inactivation was temperature-dependent above 12°C.

Initial studies of extracts of the frozen tentacles of the sea wasp, *Chironex fleckeri*, indicated that the extracts had lethal, necrotizing and haemolytic properties (Southcott and Kingston, 1959). However, it remained for Barnes (1967a) to isolate a *Chironex* nematocyst crude toxin, as Lane (1960) had done for *P. physalis* for more definitive studies. This was accomplished by an ingenious method of using human amnion membrane and electrically stimulating the tentacles. In his initial study, Barnes found that the undiluted toxin was lethal to mice at the 0·005 ml/kg body weight level but it was not known what this might be in dry weight, mg protein or protein nitrogen.

Using a modification of the extraction methods proposed by Phillips (1956) and by Lane and Dodge (1958), Endean *et al.* (1969) identified five types of nematocysts in the tentacles of *C. fleckeri*. They found proteins, carbohydrates, cystine-containing compounds and 3-indolyl derivatives in all five types. Saline extracts of the contents of the nematocysts were highly toxic to prawn and fish, and were lethal to mice and rats. In mice, the intravenous $LD_{50}$ was between 20 000 and 25 000 nematocysts, while in rats the $LD_{50}$ was approximately 150 000 nematocysts. Extracts of ruptured nematocysts elicited a strong contraction in barnacle striated muscle and in the skeletal, respiratory and extravascular smooth muscle of the rat. The contracture was sustained for varying periods, then the muscle became paralysed in the relaxed state. When the perfused heart of the toad and the exposed heart of the rat were subjected to the nematocyst extract there was a progressive failure in relaxation during succeeding cardiac cycles and the heart became paralysed in systole. Prior exposure of rat diaphragm musculature to D-tubocurarine did not modify the response to the toxin and conduction in the toad sciatic nerve was unaffected by prolonged exposure to the toxin.

Using extracts of the tentacles partially purified by Sephadex gel filtration, Freeman and Turner (1969) performed a number of pharmacological tests in mammals and concluded that the extracts produced respiratory arrest, which they attributed to central origin. They also implicated deleterious cardiac changes leading to an atrioventricular block. Blood pressure and chemistry changes were consistent with a reduced circulating blood volume and hypoxia. The toxins had a non-specific lytic effect on cells and no particular differential effect on the guinea-pig diaphragm preparation. It was found that 0·1 ml of a 5000-fold dilution of the tentacle extract would kill a 20 g mouse in less than 2 min and that the toxin was haemolytic. The fraction was nondialysable and was eluted over a range from bovine serum albumin to a molecular weight of 8000.

Crone and Keen (1969), using tentacle extracts described by Freeman and Turner (1969), obtained two toxic proteins by ion exchange chromatography on carboxymethyl and diethylaminoethyl cellulose, and by exclusion chromatography on Sephadex G-75 and G-200. The haemolytic activity was related to a protein component with a molecular weight of approximately 70 000. The second toxin had a molecular weight of about 150 000, and while both components had cardiotoxic activity the larger fraction had considerably more than the smaller fraction. The studies indicated that there was only one haemolytic fraction in the extract. This haemolysin was labile at room temperature and there was a non-linear relationship between the rate of haemolysis and dilution of the sample that was dependent on temperature and pH. Activity was inhibited by sucrose and plasma and accelerated by benzene. There was a correlation between the haemolysis titre and the $LD_{50}$ (Keen and Crone, 1969).

The cardiovascular effects of two *Chironex* tentacle extracts were studied by Freeman and Turner (1971), who found that both the fractions previously isolated produced an initial increase in systemic arterial pressure followed by a fall in pressure, bradycardia and cardiac arrhythmia. In the perfused guinea-pig heart, both toxins caused a reduction in rate, amplitude of contraction and coronary flow. The authors concluded that the cardiovascular effects were due to "direct vasoconstriction, cardiotoxicity, a baroreceptor stimulation and possible depression of the vasomotor centre".

From these various studies on *C. fleckeri*, it became apparent that the toxic material present in the nematocysts and that present in the tentacles devoid of nematocysts was lethal to mice and rats but that these materials possess markedly different biological activities. It was noted that:

Cognisance should be taken of the presence in *C. fleckeri* tentacles of toxic material which is not localized in nematocysts, when studies are contemplated of the toxic material normally injected via nematocysts into prospective prey or humans who come into contact with the tentacles of the jellyfish. Moreover, the walls of the nematocysts of *C. fleckeri* are extremely resistant to mechanical abrasion and prolonged grinding in an electrically driven microgrinder is necessary to bring about rupture of the majority of nematocysts (Endean *et al.*, 1969). Hence it is unlikely that mincing or a brief homogenization of the tentacles of *C. fleckeri* would rupture a significant number of nematocysts. Because of the method of preparation of toxic extracts adopted by Freeman and Turner (1969) it is possible that these authors studied the activity of toxic material located in the tentacles of *C. fleckeri* but which may not be present in the nematocysts . . . . Because both toxic materials have hemolytic properties it is not immediately apparent which material was studied by Keen and Crone (1969a) and by Crone and Keen (1969) but the method of preparation adopted by these authors would suggest that they studied primarily a toxic material from the tentacles which is not localized in the nematocysts. On the other hand, toxic material prepared in a similar fashion by Keen and Crone (1969b) showed dermonecrotic activity which is also shown by toxic material localized in the nematocysts but not by toxic material from tentacles devoid of nematocysts. Also, Freeman and Turner (1969) and Keen and Crone (1969a,b) state that the pharmacological properties of the extracts of whole tentacle which they had prepared were identical with those obtained from material collected by discharge of nematocysts through a human amniotic membrane. (Russell, 1971a).

There is little doubt in this writer's mind that there are chemical and pharmacological differences between extracts taken from nematocysts and those taken from tentacular tissues that are devoid of functional nematocysts. Certainly, the several methods for preparing a toxin, as suggested by Lane (1960), Barnes (1967a), Endean *et al.* (1969), Crone and Keen (1969), Freeman and Turner (1969), Burnett and Goldner (1969) and Rice and Powell (1970), do seem to lead to different products.

Endean and Noble (1971) sought to clarify some of these reported discrepancies. Using methods previously described by Endean *et al.* (1969) and Endean and Henderson (1969), they separated material within the nematocysts from the residual tentacular material. The two products were studied by injection into mice and rats, on the barnacle muscle preparation, the rat phrenic nerve-diaphragm preparation, the toad sciatic nerve-gastrocnemius and sciatic nerve conduction preparations, rat ilea and heart preparations, and certain other isolated tissue preparations.

It was found that following removal of the nematocysts from the tentacles the remaining product possessed quite different biological activities from that extracted from the nematocysts. These findings have been summarized in Table VI. On the basis of these results the authors concluded:

It is possible that the pharmacological properties of toxic material present inside the nematocysts of *C. fleckeri* are modified during discharge, during subsequent standing or storage, or during the extraction procedures adopted by the different authors who have studied toxic material from this jellyfish. The toxic material found in tentacles devoid of nematocysts could represent material which had been discharged previously from nematocysts and adsorbed on the tentacles and which had undergone a change in its pharmacological properties. However, the weight of evidence indicates that the substance of the tentacle normally contains toxic material different from that present in nematocysts. (Endean and Noble, 1971.)

TABLE VI. DIFFERENCES IN PHARMACOLOGICAL ACTIVITIES WITH EXTRACTS FROM TENTACLES AND NEMATOCYSTS OF *Chironex fleckeri*

| Preparation | Tentacle extract | Nematocyst extract |
|---|---|---|
| Muscle, barnacle | Unaffected | Affected |
| Heart, rat | Early hypotension | Early hypertension |
| Nerve, rat | Block | No block |
| Skin, rat | Very slight necrosis | Necrosis |
| Blood cells, rat | Haemolysis | Haemolysis |

Although the work of Endean and Noble (1971) indicates certain differences, it is difficult to define these accurately in the absence of a standard tissue weight or solution. Comparing a solution containing 25 000 nematocysts with a tentacular solution of 0·10 ml leaves a good deal of room for speculation. It is a pity that the various investigators have not sought a common point of comparison based on protein nitrogen content or some other such accurate measure.

Also, in this writer's experience some of the differences demonstrated by the various investigators for *Chironex* toxins (as well as several other cnidarian toxins) might be due to the dose relationships rather than the toxins employed. This seems particularly true for the rat or guinea-pig diaphragm-phrenic nerve preparation. Changes in the dose of toxins will not only cause quantitative changes but also qualitative ones. Further, when the muscle of a mammalian nerve-muscle preparation is shortened, it is most difficult to determine the effect on the indirectly elicited contraction. In fact, in a shortened muscle it is even difficult to determine the significance of the direct effect. One may speculate but direct comparisons are questionable, at best. Even adding distilled water to the bath will change the resting tension of a fibre to the extent that both qualitative and quantitative alterations may occur. Certainly, a substance that is highly irritating to an entire muscle membrane presents a problem when one hopes to define its action on the nerve, or even on the activity of a single muscle fibre.

Baxter and his colleagues at the Commonwealth Serum Laboratories in Australia, using saline extracts of nematocysts obtained after the method of Barnes (1967a), or as described by Endean *et al.* (1969), demonstrated that the biologically active fractions causing lethal, haemolytic and dermonecrotizing reactions were in the 10 000–30 000 molecular weight range. The activities were minimized by heat, formalin and EDTA, and were stabilized by peptone. There was thought to be no histamine-releasing action in the dermonecrotizing factor, and the haemolysin was not a phospholipase (Baxter and Marr, 1969).

In rabbits, a lethal intravenous dose of the venom caused laboured and deep respirations followed within several minutes by prostration, hyperextension of the head, "spasms", and respiratory and cardiac arrest. In sheep, a similar respiratory deficit was seen, followed by unsteadiness, muscle tremors and "spasms". The head drooped to one side, prostration occurred, and the tongue was said to be paralysed and cyanotic. The animal lay on its side kicking and gasping, eye reflexes were lost and there was a generalized body tremor. Death occurred within a few minutes. In primates, the animal became inactive, "dull", confused, and exhibited slight ataxia and incoordination. The eye lids drooped, the mouth was open and the head "weaved". Heart rate was irregular, and breathing became laboured, deep and irregular.

Within a few minutes the animal collapsed, the heart rate deteriorated and cyanosis developed just prior to death. Post-mortem examination revealed marked congestion of the vessels of the lungs, pulmonary oedema, the right ventricle was engorged, the kidneys and liver were congested, and the vessels of the meninges of the cerebrum were engorged (Baxter *et al.*, 1972).

Early studies on the preparation of a *Chironex fleckeri* antivenin are difficult to understand. In 1974, Baxter and Marr found that antivenin prepared against *C. fleckeri* venom neutralized extracts of *Chiropsalmus quadrigatus* and an unidentified tropical cubomedusan *in vitro* but not those of *Physalia physalis* or *Chrysaora quinquecirrha* (Desor). Again, the data are difficult to interpret because the usual standard chemical units such as the dry weight of the toxin, protein nitrogen, etc., were not used. Nevertheless the conclusions seem valid. Subsequently, a sea wasp toxoid and an antivenin were prepared (Baxter and Marr, 1975). However, preparation of the toxoid was abandoned for it was found that there was only a remote chance of an immunized person being stung, the period of effective protection was short, and there was the possibility of sensitizing individuals (Sutherland, personal correspondence, 1981). The antivenin, however, has been found very effective for *C. fleckeri* stings (Sutherland, 1979).

Considerable study has been done on the nematocyst toxin of the sea nettle *Chrysaora quinquecirrha* by Burnett and his group at the University of Maryland School of Medicine. Four distinct types of nematocysts have been observed: homotrichous anisorhiza, homotrichous isorhizas, a smaller isorhiza and the most abundant type, the microbasic eurytele (Burnett *et al.*, 1968; Blanquet, 1972). Using the nematocyst isolation method of Rice and Powell (1970), Blanquet (1972) found that the toxin was contained within the nematocyst and that the discharged nematocyst capsules and threads were free of toxin.

The crude toxin was heat labile and toxicity was abolished following boiling for 5 min. However, the toxin could be lyophilized or frozen and stored at $-20°C$ for at least a month without a significant loss of activity. Toxicity was lost following incubation for 1 h with trypsin or after protein precipitation with 10 % trichloroacetic acid. Further studies showed that the toxic material was associated with a protein fraction having a molecular weight $> 100\,000$ which could be separated into two major fractions. The more toxic of these two proteins was found to be rich in aspartic and glutamic acids, which comprised approximately 27 % of the total detectable amino acid content.

Burnett and Carlton (1977) have reviewed the cardiotoxic, dermonecrotic, musculotoxic and neurotoxic properties of *C. quinquecirrha* venom. Neéman *et al.* (1980a,b) demonstrated that the toxin, separated after a modified method of Burnett *et al.* (1968), produced striking cytological changes, including nuclear alterations and dissolution of intercellular collagen. The lethal property is thought to exert its effect by altering the transport of calcium across the conduction system of the heart (Kleinhaus et al., 1973). From experiments on rat and frog nerves and muscles, and the neurons of *Aplysia californica*, Warnick et al. (1981) concluded that the toxin appeared to induce a non-specific membrane depolarization by a sodium-dependent tetrodotoxin-insensitive mechanism which secondarily increases Ca influx; that $Ca^{++}$ does not play a direct role in the depolarization caused by the toxin; that the action is not on voltage-sensitive $Na^+$ and $K^+$ channels; and that it is not dependent upon lipase activity. The data seem consistent, although some workers might question the authors' conclusions.

Lal et al. (1981) isolated the collagenase from C. *quinquecirrha* and purified it 237-fold, using three sequential ion exchange chromatographic steps. It was activated by $Ca^{++}$ and $Na^+$, reversibly inhibited by EDTA, uninhibited by phenylmethyl-sulphonyl fluoride but inhibited by a number of chelating agents, cystine and some metal ions. A monoclonal antibody to the lethal factor of the venom has been prepared, using the crude venom as antigen. Cloned hybridomas were selected by ELISA, using the partially purified lethal factor. Ascites fluid from the cloned hybridoma-breeding mouse showed an ELISA titre of 12 800 and neutralized $2 \times LD_{50}$ of an intravenous injection of the crude venom following incubation overnight (Guar et al., 1981).

Among the sea anemones, *Anemonia sulcata* is of particular interest to man since it is of considerable medical importance in the Adriatic Sea, where it inflicts numerous stings on bathers (Maretić, 1975; Maretić and Russell, 1983). In 1973, Novak et al., obtained a partially purified, toxic, basic polypeptide from *A. sulcata* by homogenizing the tentacles, freezing the homogenate in liquid nitrogen and then thawing it repeatedly to break down the nematocysts. The extract was fractionally precipitated with ammonium sulphate, chromatographed on DEAE-cellulose and Sephadex G-50. Its molecular weight was estimated to be 6000; it gave four bands on disc electrophoresis. Its $LD_{100}$ in rats was 6 mg/kg body weight.

In 1974, Ferlan and Lebez isolated a highly basic protein toxin from *Actinia equina* by autolysing a suspension of nematocysts from the tentacles, passing it through a thick nylon screen to obtain the undischarged nematocysts, freezing and thawing the nematocysts, homogenizing and centrifuging the product, and then precipitating with acetone. The precipitate was then chromatographed on Sephadex G-50 and an electrophoretically homogenous protein with a molecular weight of 20 000, with an isoelectric point of 12·5 was obtained. The toxin, called equinatoxin, had 147 amino acid residues (Ferlan and Lebez, 1976). In rats, equinatoxin had an intravenous $LD_{50}$ of 33 μg/kg body weight and was found to have haemolytic, antigenic, cardiotropic and certain other activities (Sket et al., 1974).

In 1975, Beress *et al.* isolated three toxic polypeptides (toxins I, II and III) from *A. sulcata* by homogenizing whole specimens with ethanol and heating to 60°C, then dialysing and exposing the extract to CM cellulose, Sephadex G-50, SP Sephadex, and Bio-Gel P-2 to separate and purify each of the toxins. Toxin I contained 45 amino acid residues, toxin II, 44, and toxin III, 24. Toxins I and II had glycine at the N-terminus, while in toxin III it was arginine. The molecular weights of all three were less than 4750. Toxins I and II had similar toxicological effects. When injected into crustaceans, fishes and mammals they produced paralysis and cardiovascular changes. When injected into crabs the toxins caused convulsion and paralysis, and were lethal at 2·0 μg/kg. They also caused paralysis and death in fishes. The intramuscular $LD_{50}$'s in the juvenile cod *Gadus morhua* were 6·9 mg/kg body weight for toxin I and 0·14 mg/kg for toxin II (Möller and Beress, 1975). Toxin III caused neurotransmitter release from rat synaptosomes, as did veratridine, but the release activity appeared to occur at different receptor structures in the membrane (Abita *et al.*, 1977).

In electrophysiological studies on crayfish neuromuscular transmission, Rathmayer *et al.* (1975) demonstrated the toxicity of toxins I and II, while Alsen and Reinberg (1976) note that they were unable to show neurotoxicity in the mouse, although specific experiments are not cited. They had found that toxin II was far more toxic than toxin I, and that it produced a positive inotropic effect on isolated electrically driven atria of the guinea-pig at concentrations of $2 \times 10^{-8}$ g/ml. At higher concentrations it caused such toxic effects as contracture and arrhythmia. On the Langendorff heart preparation, low concentrations enhanced the contractile force of the atrium and ventricle, while high concentrations, $1 \times 10^{-7}$ g/ml, caused contracture and arrhythmia which appeared to be limited to the atrium. The atria proved to be much more sensitive to the toxin than the ventricular muscle. The authors noted the differences between the toxin and certain cardiac glucosides.

Ferlan (1977) found that, *in vitro*, equinatoxin exhibited strong lytic action on erythrocytes; it did not have phospholipase activity. He also studied the effects of different ions on the toxin's haemolytic activity. Giraldi *et al.* (1976) demonstrated that a suspension of phospholipids would reduce haemolytic activity. Penčar *et al.* (1975) demonstrated that a greater concentration of the toxin was required to cause changes in the molecular order of the membrane than that required for complete haemolysis. Maček and Lebez (1981) showed that $1 \times 10^{-10}$ M of the toxin caused complete haemolysis of sheep erythrocytes, while $Ca^{++}$, $Mg^{++}$ and $Ba^{++}$, in that order, stimulated the rate of haemolysis. EDTA decreased haemolytic activity by 30%, while $Sr^{++}$ and $Mn^{++}$ had no effect. The authors concluded that the data do not clarify the role these ions play in equinatoxin-induced haemolysis and suggest that the lytic activity is accelerated following binding of the toxin to the membrane "due to changing physical-chemical characteristics of membrane phospholipids" (Giraldi *et al.*, 1976).

A cytolytic toxin, metridiolysin, from the anemone *Metridium senile* has similar haemolytic activity, inhibited by cholesterol (Bernheimer and Avigad, 1978). Like equinatoxin, haemolytic activity is restricted to a relatively narrow pH range, with an optimum at 5·6. The optimum is 8·8, and 8·5 in the presence of Ca$^{++}$. Equinatoxin exhibits no haemolytic activity below a pH value of 6·5.

In 1964, Munro demonstrated that tentacle homogenates of the large Caribbean anemone *Condylactis gigantea* had a marked paralytic effect on crustaceans. Taking advantage of this observation, Shapiro and his colleagues at Harvard carried out a series of chemical and pharmacological studies on a stable acetone powder from tentacle homogenates of the species (Shapiro, 1968a,b; Shapiro and Lilleheil, 1969). The tentacles were washed, blotted, cooled, homogenized and centrifuged at 10 000 rpm for 1 h. Cold acetone was added to the supernatant, the suspension filtered and washed with acetone and then dried. The uniform, fine yellow powder was stored at —15°C under vacuum. It was then resuspended in cold buffer, homogenized and centrifuged at 60 000 rpm for 60 min. The total protein in the solution was 45 % of the dry powder. The powder was found to retain its biological activities over a 14-month test period. Gel filtration was done on Sephadex G-50 and G-75 and ion-exchange chromatography was carried out on Cellex D. Behaviour suggested the toxin was a basic protein with a molecular weight of 10 000–15 000. It was removed from DEAE-cellulose at basic pH and low ionic strength.

Assays were performed with the acetone powder on the crayfish *Orconectes virilis*. On envenomation, the crayfish exhibited a paralysis characterized by an initial rigid or spastic phase, followed by a flaccid phase. Impaired reflexes were a constant finding. The righting reflex was used as a criterion for determining the degree of paralysis. Specimens which exhibited such paralysis died within 48 h. The immobilization dose ($ID_{50}$) for the acetone powder in crayfish was about 1 μg/kg body weight and the yield from a 70 g anemone with 23 g of tentacles was 1 g of the acetone powder, or a sufficient amount to paralyse approximately 2100 kg crayfish. It was found that the acetone powder solutions gave results which were equal to freshly prepared tentacle homogenates. Homogenates of non-tentacular tissue did not show significantly paralytic activity.

The acetone powder solution lost 100 % of its activity on heating to 100°C for 30 min. Pronase destroyed the activity; trypsin destroyed 80 % of the activity after 15 h incubation at 23°C. The active component was non-dialysable. Gel filtration of the acetone powder solution on G-50 showed that the activity was contained in a single symmetrical peak at 2·12 times the void volume. The average peak volume was 2·16 times the void volume and recovery was 85–90 %. Similar runs with crude tentacle homogenates on G-75 give almost identical results. Toxic activity did not coincide with any peak in the total protein. The active portion was 75 times more potent than the acetone powder solution. It was found that the contaminating proteins were lighter than the active fraction and that they could be separated off on Sephadex G-50. Initial chromatography on G-50 followed by chromatography on G-75 increased the purification of the acetone powder approximately 150-fold.

With ion-exchange chromatography on Cellex D, the recovery of activity was dependent on the conditions of attachment. When G-50 purified acetone powder was loaded at $\gamma/2$ (0·0225), the toxic activity came off as a sharp band between ionic strengths of 0·062 and 0·088. The increase in specific activity was 21-fold greater than the G-50 purified acetone powder. Stepwise changes in ionic strength were found to be more reliable: loading acetone powder or G-50 purified acetone powder at an ionic strength of 0·06 and removing it at 0·08 gave yields of 75 %, and purification of 25- to 27-fold.

Purification using both gel filtration and ion exchange gave a product with an immobilization dose ($ID_{50}$) of approximately 1 μg protein/kg crayfish. This purified toxin was non-dialysable, destroyed by pronase, and had a half-life in phosphate buffer of 24 h at 4°C. It was markedly inactivated by shaking and foaming, and lost all activity within 10 min at 100°C. The behaviour of the toxin, estimated from the rate of migration on columns indicated that it was a basic protein with a molecular weight of 10 000–15 000.

Further studies on a crayfish preparation showed that the toxin had a direct effect on the crustacean nerve but not on the muscle membrane. No evidence for a truly synaptic effect was observed. The toxin caused the motor axons to fire repetitively in high frequency bursts or low frequency trains. The spastic phase of paralysis was followed by a flaccid phase attributed, in part, to a neural conduction block. Using the lobster giant axon and crayfish slow-adapting preparations it was found that the toxin transformed action potentials into prolonged plateau potentials of up to several seconds duration, and that the eventual conduction block was not due solely to depolarization. It was suggested that the plateaus were caused, at least in part, by a prolonged membrane permeability following the initial excitation.

In 1973, Turlapaty et al. described a central nervous system stimulant in the form of a basic polypeptide isolated from homogenized tissues of the anemone *Stoichactis kenti*. The stimulation was described as "fighting episodes" and the authors suggested that the action was due to a central adrenergic mechanism. Devlin (1974) studied a partially purified toxin from *S. helianthus* which had a mouse intraperitoneal $LD_{50}$ of 0·25 mg/kg and possessed haemolytic properties inhibited by sphingomyelin (Bernheimer and Avigad, 1976). Michaels (1979) and Shin et al. (1979) have shown that the cytotoxic poison from the anemone *Stoichactis helianthus* acts upon black lipid membranes and liposomes by channel formation and detergent action. This mechanism is also suspected for the haemolytic activity of some of the other sea anemone "cytolytic" toxins. However, there appears to be a considerable number of different cytolytic toxins in the sea anemones and the mechanisms by which they damage membranes may be quite different. Hessinger and Lenhoff (1976), for instance, demonstrated that the toxin of *Aiptasia pallida* caused lysis through the action of phospholipase A on membrane phospholipids, while other workers suspect ionic or non-enzymatic roles.

Mebs and Gebauer (1980), in a well-executed group of experiments on a *Stoichactis* species, separated biologically active polypeptides by gel filtration and ion-exchange chromatography. Following gel filtration on Sephadex G-75, two toxins were obtained, one having haemolytic activity and the other causing death in mice and crabs, and inhibiting the proteolytic activity to trypsin and chymotrypsin. The haemolysin was further purified by chromatography and separated from the proteinase-inhibiting activity, which was resolved into three inhibitors. The molecular weight of the haemolysin was approximately 10 000; it had no phospholipase A activity nor was lecithin required for haemolysis. Complete haemolysis of a 0·6% human erythrocyte suspension was achieved at a concentration of 0·86 μg/ml. The haemolysin killed 1 cm fish (*Poecilia reticulata*) within 1 h at a concentration of 4–5 μg/ml tank water. It was not deleterious to mice at 10 mg/kg body weight.

The lethal mouse activity was eluted from columns with proteinase inhibitors and had a suggested molecular weight of 6000. Although Mebs and Gebauer were unable to perform definitive lethality determinations in mice or crabs, they did find a subcutaneous minimum lethal dose of approximately 2·3 mg/kg for mice and an intramuscular minimum lethal dose of about 0·5 mg/kg for crabs. The amino acid composition of inhibitors 2 and 3, calculated on a basis of a molecular weight of 6000, gave a weight of 5800 for the main fraction, inhibitor 2. The inhibitors lacked tryptophane, while inhibitor 2 had four half-cystinyl residues and inhibitor 3 had six residues. The terminus was masked.

Several other Actiniidae species have been shown to contain toxins. Norton *et al.* (1976) isolated a polypeptide, termed anthopleurin-A, from *Anthopleura xanthogrammica* that closely resembles toxin II from *A. sulcata* in its amino acid sequence (Tanaka *et al.*, 1977). In mice, anthopleurin-A had an $LD_{50}$ of 0·3–0·4 mg/kg body weight and was said to stimulate cardiac activity. A second polypeptide toxin, anthopleurin-B, has also been identified in the same anemone.

Little is known of the toxic properties of the alcyonarians. In 1974, Neéman *et al.* observed that certain soft corals found in the Red Sea were not attacked by predators. One such species was *Sarcophyton glaucum* (Quoy & Gaimard). It was observed that fish would die if they were kept in the same container with this coral. This alcyonarian was extracted with hexane in a Soxhlet over a 24-h period. After cooling the extract, a white crystalline precipitate appeared. The crystals were isolated, dried, and after recrystallization from acetone-hexane their melting point was found to be 133–134°C and ($\alpha_D^{25}$ + 92°C, 1·0, $CHCl_3$). NMR were taken using a solution with TMS as an internal standard. Mass spectra were done. Figure 7 shows a more recent structure of the toxin, called sarcophine. The formula is $C_{20}H_{28}O_3$. The toxin contains three methyl groups, an additional methyl on an epoxide bearing carbon, as well as three particular down-field appearing protons. The $LD_{50}$ in the fish *Gambusia affinis* was 3 mg/l of water and the first deleterious signs were excitation, turning and jumping. This was followed by prostration, sinking to the bottom and cessation of gill cover movement. In rodents, subcutaneous injections caused excitation, increased respirations, and "paralysis" followed by coma and death. When rats were fed sarcophine there was a decrease in cardiac, respiratory and motor activity. On the isolated guinea-pig ileum, 0·2–0·8 mg of the toxin/l bath solution produced a strong anticholinesterase response. The authors suggested that materials like sarcophine form part of the mucus secreted by the coral and act as a repellent. Erman and Neéman (1977) found that sarcophine inhibited the enzyme phosphofructokinase. Linear uncompetitive inhibition of the enzyme was thought to be due to the reaction of the sarcophine with thiol groups of the enzyme.

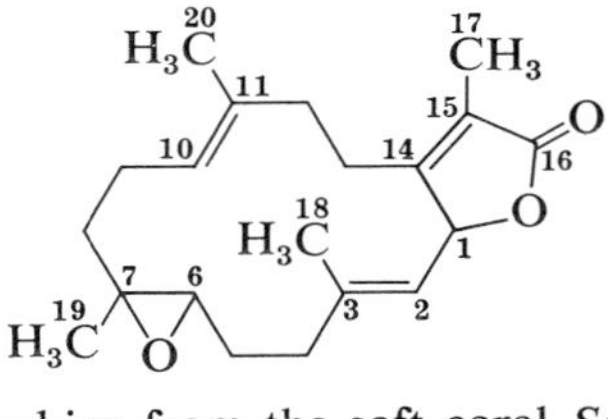

FIG. 7. Structure of sarcophine from the soft coral *Sarcophyton glaucum* (Erman and Neéman, 1977).

In 1967, Hashimoto and colleagues collected specimens of the file fish, *Alutera scripta*, from the Ryuku Islands following a report that several pigs had died following their eating of the viscera of this fish. On examining the gut contents of the fish the investigators found polyps of the zoanthid *Palythoa tuberculosa* (Hashimoto *et al.*, 1969a). About the same time, Scheuer and his group were investigating the toxin from *Palythoa toxica*, found at Hana off the island of Maui and called *lima-make-O-Hana* (death seaweed of Hana). Dr. Banner tells the story of how he and Dr. Helfrich first went in search of the organism on Maui in 1961 (where they found that the culprit was not an alga but a zoanthid). Much superstition was associated with the organism and "anyone who collected *limu-make-O-Hana* will meet with disaster," so believed the native islanders. When Banner and Helfrich returned to their laboratory on Coconut Island that evening they found that the laboratory building housing their facility had burned to the ground. Needless to say the islanders entertained their own aetiology.

Palythoa toxin, palytoxin, was first isolated by Moore and Scheuer in 1971. It was found to be a most potent poison, having a mouse intravenous $LD_{50}$ of 0·15 µg/kg body weight. It has an approximate molecular weight of 3000 and appears to have a unique structure which, as yet, is undetermined. Attaway and Cierszko (1974) reported chills and fever after grinding dried specimens of *P. caribaeorum*, which were being studied for the presence of wax esters. They isolated a toxin which appeared identical to palytoxin. They also found other toxic zoanthids in various parts of the Pacific. Accidental contact with the mucus of *Palythoa* through the abraded skin is said to produce weakness and malaise, as well as localized irritation.

The toxin from *P. vestitus* Verrill was found to have an intravenous $LD_{50}$ that varied from 0·033 to 0·450 µg/kg depending on the test mammal used. It was non-toxic by oral and rectal routes but caused marked irritation to the skin and eyes, and on injection it caused haemorrhage and necrosis (Wiles *et al.*, 1974). Further studies by Moore and his group (see Moore and Bartolini, 1981) and by Hirata *et al.* (1979) and Uemura *et al.* (1980) have shown that palytoxin contains several different units, depending on the species. The formulas were found to be $C_{129}H_{221}N_3O_{54}$ (Mr 2659) for palytoxin from the Tahitian *Palythoa* sp. and $C_{129}H_{223}O_{54}$ (Mr 2677) for the two palytoxins from *P. toxica* (Moore and Bartolini, 1981). These findings are consistent with the molecular weights reported by Macfarlane *et al.* (1980) for the two components in palytoxin from the Okinawan cnidarian, *P. tuberculosa*.

Although some ciguateric fishes are coral feeders, and several authors have implicated corals as a major factor in ciguatera poisoning, Hashimoto and Ashida (1973) were unable to demonstrate ciguatoxin, a major toxin involved in ciguatera poison, as a component of the coral samples they studied. However, they did find several water-soluble toxic fractions in the corals, *Goniopora* spp., *Palythoa tuberculosa*, *Coeloseris mayeri*, *Clavularia* sp., *Acropora* sp., *Cyphastrea* sp. and *Pavona obtusata*.

*Goniopora* toxin was prepared as shown in Fig. 8. It was found to be rich in aspartic acid, proline, isoleucine, leucine and lysine. The authors suggested a molecular weight of 12 000. Its intraperitoneal minimal lethal dose in mice was 0·3 mg/kg body weight. The mice exhibited hypoactivity, followed by rigor, respiratory distress, paralysis and cyanosis. Skeletal muscle rigidity just prior to death was a characteristic finding. A 0·9 saline extract of the coral had a haemolytic effect on rabbit erythrocytes and when purified on Sephadex G-50, followed by chromatography on CM Sephadex C-50 its haemolytic activity was increased to that near Merck saponin. It had a suspected molecular weight of 30 000. On disc electrophoresis it still showed three bands (Hashimoto, 1979).

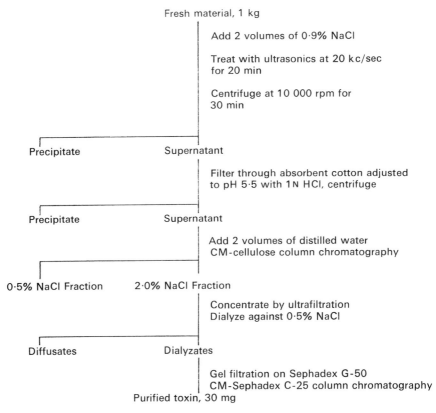

Fresh material, 1 kg

Add 2 volumes of 0·9% NaCl

Treat with ultrasonics at 20 kc/sec for 20 min

Centrifuge at 10 000 rpm for 30 min

Precipitate          Supernatant

Filter through absorbent cotton adjusted to pH 5·5 with 1 N HCl, centrifuge

Precipitate          Supernatant

Add 2 volumes of distilled water CM-cellulose column chromatography

0·5% NaCl Fraction        2·0% NaCl Fraction

Concentrate by ultrafiltration Dialyze against 0·5% NaCl

Diffusates          Dialyzates

Gel filtration on Sephadex G-50 CM-Sephadex C-25 column chromatography
Purified toxin, 30 mg

Fig. 8. Purification scheme for the toxin from *Goniopora* spp. (Hashimoto, 1979).

## C.  *Clinical Problem*

As pointed out by Halstead (1965), it has long been known that the nemato-cysts of certain cnidarians can penetrate the human skin. While most nemato-cysts are capable of piercing only the thin membranes of the mouth or con-junctiva, some possess sufficient force to pierce the skin of the inner sides of the arms, legs and more tender areas of the body. Still others can penetrate the thicker skin of the hands, arms and feet (Russell, 1965a,b). Swallowing the tentacles or even the umbrella can cause epigastric pain and discomfort (Russell, 1965a,b; Halstead, 1965; Hashimoto, 1979; Maretić and Russell, 1983). Although stingings are usually associated with contact with the tentacles, Fig. 9 demonstrates that effective nematocyst discharge can occur from the bell (Maretić *et al.*, 1980). In 1965, approximately 70 of the 9000 species of cnidarians were noted to have been involved in injuries to man (Russell, 1965a,b). More recent clinical records would indicate that about 78 species have been implicated in such injuries.

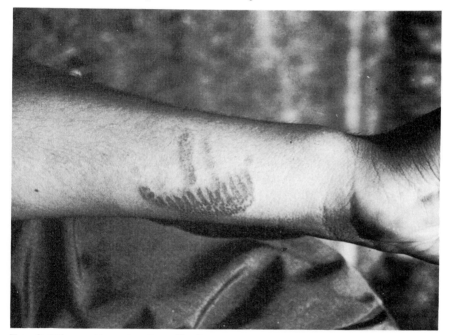

Fig. 9. Sting pattern showing effective envenomation from the bell as well as the tentacles following contact with a jellyfish.

The cutaneous lesions, as well as other clinical manifestations produced by the various cnidarians vary considerably, depending on the species involved and the number of fired nematocysts. Contrary to common belief, the stings of many cnidarians produce little or no immediate pain. Sometimes, itching is the first complaint that calls the victim's attention to the injured area, and this may not be for hours following the initial contact. In the author's experience, stings by hydroids usually do not produce pain, although there may be subsequent localized discomfort. In most cases, the lesions produced by hydroids are minimal. The fire or stinging corals, *Millepora*, produce small reddened, somewhat papular eruptions, which appear 1–10 h after contact and usually subside within 24–96 h. In severe cases the papules may proceed to pustular lesions and subsequent desquamation. The stinging is usually associated with some localized, pricking-like pain, generally of short duration, and with some subsequent pruritus and minimal swelling (Russell, 1965a,b, 1971a).

Contact with the Portuguese man-of-war, *Physalia*, causes immediate pain, sometimes severe, and the early appearance of small reddened, linear, papular eruptions. At first the papules are surrounded by an erythematous zone but as their size increases the area takes on the appearance of an inflammatory reaction with small periodic, demarcated, haemorrhagic papules. In some cases these papules are very close together, indicating multiple discharge of nematocysts as the tentacle passed over the injured part. The papules develop rapidly and often increase in size during the first hour. The affected area becomes painful and severe pruritis is not uncommon. Pain may spread to the larger muscle masses in the involved extremity or even to the whole body. Pain sometimes involves the regional lymph nodes. In some cases the papules proceed to vesiculation, pustulation and desquamation. I have seen several cases in which hyperpigmentation of the lesions was obvious for years following a stinging (Russell, 1965a,b).

General systemic manifestations may also develop following *Physalia physalis* envenomation. Weakness, nausea, anxiety, headache, spasms in the large muscle masses of the abdomen and back, vascular spasms, lachrymation and nasal discharge, increased perspiration, vertigo, haemolysis, difficulty and pain on respiration, described as being unable to "catch one's breath", cyanosis, renal failure and shock have all been reported (Russell, 1965a,b, 1966b; Drury *et al.*, 1980; H. A. Guess, unpublished communication, 1982).

Contact with most of the true jellyfishes gives rise, in the less severe cases, to manifestations similar to those noted for *Physalia*, with symptoms sometimes disappearing within 10 h. In the more severe cases there is immediate, intense, burning pain, with contact areas appearing as swollen wheals, sometimes purplish, and often bearing haemorrhagic papules. The areas may proceed to vesiculation and necrosis. Localized oedema is common and in the more severe cases muscle mass pain, difficulties in respiration, and severe spasms of the back and abdomen with vomiting are reported. Vertigo, mental confusion, changes in heart rate, and shock are sometimes seen (Barnes, 1960; Russell, 1965a,b).

The sea wasps, *Chironex fleckeri* and *Chiropsalmus quadrigatus* and certain other species are extremely dangerous cubomedusae responsible for a number of deaths, particularly in Australian waters. The former, the more dangerous, may have a bell of 15 cm across with tentacles, when their total

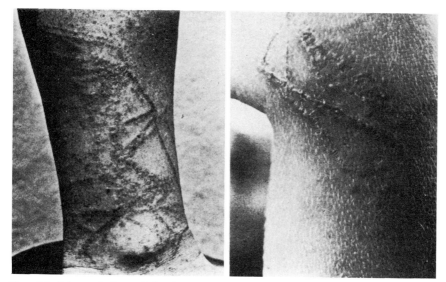

Fig. 10. Linear lesions following stinging by *Chironex fleckeri* (Courtesy Jack Barnes).

length is added together, of over 90 m (Barnes, 1960). Although systemic effects usually develop within 5–150 min following envenomation (Barnes notes an average onset of 20 min), some deaths occur in less than 5 min. Stings by these cubomedusae cause a sharp prickling or burning sensation with the appearance of a wheal, which at first appears like a "rounded area of gooseflesh". An erythematous wheal soon develops and may become considerably larger than the area of contact. At first, it may show little pattern to suggest whether the stinging had been by tentacles or by the animal's umbrella. The wheals may either disappear, as when the stinging is minimal, or after an hour or so become enlarged as the nematocyst punctures become more apparent and appear as very small haemorrhagic vesicles surrounded by inflammation. A stinging pain develops and may persist for 1–3 h. In linear lesions the nematocyst injuries may be no more than 5 mm wide but extend for 10 cm or more (Fig. 10). Where the stingings have been from the umbrella the lesions appear as a cluster or an oedematous wheal. Vesiculation and pustular formation may occur and full thickness skin necrosis is not uncommon. Oedema about the area may persist for ten or more days (Barnes, 1967a,b, personal communication, 1967; and by author).

Some distinguishing characteristics of stings by Australian cnidarians are shown in Table VII.

TABLE VII. CHARACTERISTICS OF MODERATELY SEVERE CNIDARIAN STINGS OFF
                                    AUSTRALIA

| Signs and symptoms | *Physalia* | Cubomedusae | *Cyanea* |
|---|---|---|---|
| Wheal | | | |
|   Type | Single or replicated line | Multiple lines | Multiple lines |
|   Width | Variable | 3–7 mm | 2–3 mm |
|   Pattern | Irregular | Transverse bars | Zig-zag |
|   Duration | 1–4 h | 2–24 h | 30–60 min |
| Pain: | | | |
|   Severity | Moderate | Marked | Less marked |
|   Duration | 30 min–2 h | Many h | 10–30 min |
| Necrosis or vesication | Rare | Usual | Rare |

Russell (1971b) and adapted from Barnes (1960).

Stingings by the anemones are usually of lesser consequence than those inflicted by the jellyfishes and rarely are they as painful or disabling. The lesion area takes on a reddened and slightly raised appearance, bearing irregularly scattered pin-head size vesicles or haemorrhagic blebs (Maretić and Russell, 1983). The area becomes painful, particularly to touch or heat. In stings by *Anemonia sulcata* seen by the author there has been some diffuse oedema around the injured site. Residual hyperpigmentation or hypopigmentation is unusual following anemone stings.

Stings by the stony corals (*Acropora*) are said to give rise to some minor pain often followed by itching and the development of small diffuse wheals which may progress to vesiculation but rarely necrosis. "Sponge fisherman's disease" is due to the actinian *Sarortia elegans*. Troublesome are small spicules of coral which sometimes break off and become embedded in the skin, occasionally giving rise to infection.

On some Pacific islands, as well as elsewhere in the world, sea anemones are eaten following cooking but some are apparently poisonous whether uncooked or cooked. *Rhodactis howesi* and *Physobrachia douglasi* are poisonous when eaten raw but said to be safe when cooked. *Radianthus paumotensis* and another *Radianthus* species are said to be poisonous, whether raw or cooked. Intoxication is typified by nausea, vomiting, abdominal pain and hypoactive reflexes. In severe cases, marked weakness, malaise, cyanosis, stupor and death have occurred (Cutress, personal communication, 1958; Martin, 1960; Farber and Leske, 1963; Hashimoto, 1979).

## V.  Platyhelminthes

The phylum Platyhelminthes, the flatworms, are characterized by having a
flat and bilaterally symmetrical, unsegmented body, and having definite
anterior and posterior ends. Some authorities note that they are the first
animals to have a head and tail end. In the class Turbellaria, the epidermis is
ciliated, while in Trematoda and Cestoda the surface is covered by a thick
cuticle. Most species of Turbellaria are free-living, benthic and marine.
Their epidermis is cellular or syncytial and has some cilia and rhabdoids.
These flatworms vary in length from about 0·5 to many centimetres. There
are approximately 10 000 species of Platyhelminthes, of which about 2000
are Turbellaria.


### A.  *Poisoning*

The turbellarians are thought to be the only group of flatworms that are
toxic. Their mechanism of poisoning is not understood. There are certain
glandular cells and rhabdoids in the animals' integument and both are
secretory, although the exact nature of their secretions is unknown. They
are thought to play a part in producing a toxin used in the animals' defensive
armament.

Arndt (1925) and Arndt and Manteufel (1925) studied crude extracts of
some freshwater flatworms, *Dendrocoelum lacteum* Müller, *Polycelis nigra*
Ehrenberg, *P. cornuta* O. Schmidt, *Planaria gonocephala* Dug., *P. lugubris*
O. Schmidt and *Bdellocephala punctata* Pall. and found that they caused a
reversible (on washing) cardiac arrest in the frog heart. Arndt (1943) also
studied crude saline extracts of some marine species, *Leptoplana tremellaris*
Oersted, *Stylochus neapolitanus* (Delle Chiaje), *Thysanozoon brocchi* Grube,
*Yungia aurantiaca* Delle Chiaje, *Bdelloura candida* (Girard) and *Procerodes
lobata* (Schmidt) and found that they were toxic to guinea-pigs and that
heating the extracts to 100°C for 1 min destroyed this toxicity.

Thompson (1965) found a strongly acetic secretion, pH 1, in the epidermis
of the worms *Cycloporus papillosus* and *Stylosatomum ellipse*. The secretion
probably acts as a repellent to those animals which might try to feed upon
the flatworm. The flatworms are of little clinical importance.


## VI.  Rhynchocoela

The phylum Rhynchocoela, the nemerteans, or ribbon worms, is a small
group of slender unsegmented worms distinguished from the Platyhelminthes

by their rounded body which usually tapers somewhat towards one or both ends, and the presence of a prominent, eversible proboscis. These animals are carnivorous, feeding on annelids, crustaceans, molluscs and even fishes. Most species are marine. They vary in length from less than a millimetre to over a metre. There are about 800 living species.

## A.  *Poisoning*

Envenomation by certain nemerteans occurs when the threadlike proboscis is extended explosively during the capture of prey or on handling. There is some question as to where the venom is produced and how it is transferred to the stylet when it is fired through the proboscis. It is known that in feeding the worm jabs its prey with its proboscis stylet and the prey becomes paralysed. To be effective this may be done several or many times. In *Cerebratulus lacteus* (Leidy) the toxin appears to be confined to peripheral tissues, principally the integument and the proboscis, both of which have dense populations of gland cells (Kem and Blumenthal, 1978). For a more detailed discussion of this problem the reader is referred to Halstead (1965) and the fine review by Kem (1973).

In 1936 and 1937, Bacq found two toxic substances in alcoholic tissue extracts of *Amphioporus lactifloreus* (Johnston), *Drepanophorus crassus* (Quatrefages) and certain other species while studying the distribution of choline esters in various animals. He observed that when the extract he called *amphiporine* was injected into the lymph glands of frogs there was an increase in respirations, dilated pupils and extension of the limbs, with recovery within 30 min. The toxin did not seem to be localized in any part of the animal's body and was not released into the surrounding water by the worm. Bacq concluded that amphiporine was an alkaloid similar to nicotine. The second substance, *nemertine*, from *Lineus lacteus* Montagu and *L. marinus* (Montagu) was said to be a "nerve stimulant" but no details of its properties were given. King (1939) attempted to isolate amphiporine from *A. lactifloreus* and carried out a number of extraction procedures but was unable to obtain a product suitable for analytical work. However, he noted that the action of amphiporine on certain pharmacological preparations was identical to that produced by nicotine.

It was not until the late 1960s that further studies were carried out on nemertean toxins. Kem (1969, 1971) and Kem *et al.* (1971) studied *Paranemertes peregrina*, *Cerebratulus lacteus* (Leidy), *L. ruber*, *L. viridis* and *A. angulatus*. Toxin was obtained by mechanical or chemical (1 % acetic acid) induction, which caused the release of copious mucus secretions. It appeared to be confined to peripheral tissues and the proboscis (Kem and Blumenthal,

1978). Quantal assays were done by measuring the median effective dose ($PD_{50}$) necessary to paralyse a 20 g crab. Paralysis was measured as the loss of righting ability. The secretions were then centrifuged, the supernatant adjusted to pH 5·0 and applied to a CM cellulose floc. The adsorbed toxin could be stabilized at a temperature of $-20°C$ or on lyophilization. Extracts were then subjected to 1 M ammonium acetate at pH 7·0 and purified on G 50F Sephadex and CM 52 cellulose column chromatography. Fractions were monitored at 280 and 230 nm, pooled and lyophilized. Assays were performed by measuring the median effective dose ($PD_{50}$) necessary to paralyse a 20 g crayfish. Paralysis was measured as the loss of righting ability.

Kem (1976) found two main types of *Cerebratulus* toxins, one of which had almost all of the mouse lethality factor, toxin A, and very little crayfish activity, and the other, toxin B, which was very toxic to crayfish but not to

TABLE VIII. AMINO ACID COMPOSITIONS OF PURIFIED *Cerebratulus lacteus* TOXINS (RESIDUES PER MOLECULE)

| Amino acid | A toxins | | | B toxins | | | |
|---|---|---|---|---|---|---|---|
| | II | III | IV | I | II | III | IV |
| Lysine | 14 | 14 | 16 | 7 | 8 | 8 | 10 |
| Histidine | 0 | 1 | 1 | 3 | 2 | 1 | 1 |
| Arginine | 3 | 3 | 4 | 3 | 2 | 3 | 3 |
| CmCystine | 6 | 6 | 8 | 6 | 8 | 8 | 8 |
| Hydroxyproline | — | — | — | — | 1 | — | 1 |
| Aspartic acid | 6 | 6 | 5 | 6 | 6 | 6 | 5 |
| Threonine | 3 | 3 | 2 | 1 | 3 | 1 | 1 |
| Serine | 7 | 8 | 6 | 1 | 4 | 2 | 1 |
| Glutamic acid | 4 | 4 | 5 | 3 | 4 | 5 | 4 |
| Proline | 4 | 2 | 5 | 1 | 0 | 0 | 0 |
| Glycine | 11 | 12 | 9 | 5 | 5 | 5 | 5 |
| Alanine | 12 | 13 | 12 | 4 | 6 | 7 | 8 |
| Valine | 5 | 7 | 5 | 3 | 1 | 0 | 0 |
| Methionine | 0 | 0 | 2 | 1 | 0 | 0 | 0 |
| Isoleucine | 7 | 6 | 6 | 1 | 1 | 3 | 3 |
| Leucine | 6 | 5 | 7 | 1 | 0 | 1 | 1 |
| Tyrosine | 1 | 1 | 2 | 1 | 2 | 2 | 2 |
| Phenylalanine | 2 | 2 | 2 | 2 | 0 | 0 | 0 |
| Tryptophan | 2 | 2 | 2 | 0 | 2 | 2 | 2 |
| Total residues | 93 | 95 | 99 | 49 | 56 | 54 | 55 |
| Formula wt | 9771 | 9799 | 10 533 | 5453 | 6043 | 5898 | 6111 |
| Molecular wt (10% Agarose) | 10 400 | 10 500 | 10 400 | 5380 | 5930 | 5970 | 6000 |
| Amino Acid differences | 10 | 0 | 23 | 23 | 9 | 0 | 7 |

From Kem, personal communication, 1982.

mice. Further studies showed that toxin A contained three toxins, which Kem and Blumenthal (1978) found to be very basic single chain polypeptides, having 94–98 amino acid residues and 3–4 disulphide bonds. On sequencing of one of these fractions, A-III, it was found that the amino-terminal half possessed most of the charged residues, as well as 4 or 6 half-cystines, while the carboxyl-terminal half was substantially hydrophobic. This toxin rapidly lysed a variety of cells. It was specifically inhibited by sphingomyelin. Studies with human erythrocytes indicated that sensitivity was roughly proportional to membrane sphingomyelin : phosphatidylcholine ratios (Kem et al., 1980). They suggested that since the three toxins had very similar amino acid compositions they are probably "isotoxins", and as preliminary amino acid sequencing findings indicated homology, and all three had potent cytolytic effects, including 50 % lysis of human red cells in the 1–5 μg/ml concentration range, the toxins may be evolutionary variants of a common ancestral A toxin. Table VIII shows the most recent amino acid composition of the toxins. It is refreshing to this writer to see this simple, applicable classification for these toxins, Cerebratulus A, A-II, A-III, etc., rather than what Hillard (1938) called the "scientific ego trip", the gimmick names so frequently found in the toxinology literature.

The manner in which Cerebratulus A toxins disrupt cell membranes has been the object of several studies and, again, it is refreshing to find Kem's statement that these toxins "should be designated as cytotoxins rather than haemolysins, as they lyse a variety of cells". I might go one step further and advise we use the word cytotoxic rather than cytotoxins, since I am rather sure, from experiences with other toxins, that we will find that the Cerebratulus toxins have pharmacological properties other than their cytotoxic ones. Indeed, Kem and Blumenthal (1978) note that sublytic concentrations of toxin A-III depolarize resting potentials and block action potentials in the squid giant axon and in the dog Purkinje cells. In the squid axon voltage clamp experiments the sodium channel was completely blocked, even when depolarization of the resting potential was prevented by hyperpolarizing currents. Further study indicated that the depolarization was not due to the selective opening of sodium channels and that it may involve a non-specific increase in membrane permeability to several ions, including $Na^+$ and $Ca^{++}$ but this has yet to be established (Kem et al., 1980).

Posner and Kem (1978) point out the pharmacological similarities on rabbit atrial myocardial cells between A-III and "cardiotoxin" of the Formosan cobra, and Kem and Blumenthal (1978) noted that sublytic concentrations of cytotoxin A-III inhibited voltage-dependent $Na^+$, $K^+$ and $Ca^{++}$ channels in the squid axon. The reader might well ask if the fundamental mechanisms responsible for the deleterious effects of "cytotoxins", "cardiotoxins" and "neurotoxins" are not basically very much the same. Perhaps we should consider a receptor mechanism rather than a receptor site.

The four *Cerebratulus* B toxins were also shown to have large portions of basic amino acids (see Table VIII). It can be seen from the table that B-III and B-IV are very similar, both have four disulphide bonds and the common absence of three amino acids. They all appear to have molecular weights of approximately 6000.

Following sequencing and further studies of B-IV, it was concluded that the Tyr-9 was directly related to the toxin–receptor interaction. B-IV prolongs repolarization of the action potential in the crayfish giant axon but this phenomenon is not seen in frog sciatic, garfish olfactory or squid axon preparations (Blumenthal and Kem, 1980). The preferential action of some venoms and venom fractions for the crayfish preparation is not without precedent (Parnas and Russell, 1967). All four toxins produce similar effects in the crayfish, namely tremor, spastic convulsions followed by flaccid paralysis, and death (Kem and Blumenthal, 1978).

# VII. Annelida

The phylum Annelida, the bristle worms, bloodworms and polychaete worms are elongated, segmented, bilaterally symmetrical worms with essentially similar segments that resemble each other in both their external appearance and internal structure. They have a nonchitinous cuticle secreted by the epidermis and containing numerous unicellular gland and sensory cells, most having chitinous setae, a complete digestive system, a closed circulatory system and, in some species, a well-developed chitinous jaw. Most of the venomous marine species are found in the order Polychaeta, which has approximately 5500 species. They have a world-wide distribution and are, for the most part, benthic; most species are creeping or burrowing but some are sedentary, and a few are pelagic.

## A. *Poisoning*

The toxicity of the marine annelids is associated with their bristle-like setae or their biting jaws. Some species contain toxic substances within their bodies. With respect to the setae, these elongated chitinous bristles project from their parapodia. Each seta is secreted by a single cell at its base. In some species the setae can be extended or retracted so that with the worm at rest they appear short and scarce but when the animal is aroused the setae are extended and, as Halstead (1965) points out, the animal "appears to be a mass of bristles".

The most often implicated genera of polychaetes involved in stingings are *Chloeia*, *Eurythoe* and *Hermodice*. The setae of *Eurythoe* and *Hermodice* are known to be hollow and at times are filled with fluid (Halstead, 1965) but the status of *Chloeia* as a venomous annelid is equivocal. Pope (1947) notes that when the spines of *E. complanata* (Pallas) break off in the skin they cause an "itching-burning sensation" which will continue as long as they remain in the skin. Cleland and Southcott (1965) note *Aphrodite australis* (Baird) as a species capable of causing injury and *H. carunculata* (Kinberg) has been incriminated by Mullin (1923), *C. flava* by Tweedie (1941), *E. aphrodite* by Pope (1963) and *H. carunculata, Amphinema brasiliensis* and *C. euglochis* by Phillips and Brady (1953)

Penner (1970) found that the setae of the bristleworm *H. carunculata* produced pain and transient numbness when they were broken off in the skin. The pain or stinging sensation persisted for 13·5 h and the numbness, which involved the entire extremity, persisted for about 30 min. He suggested that the setae contained a "neurotoxin", which was emptied into the wound when the bristles broke off.

The marine annelida *Lumbriconereis heteropoda* was found by Nitta (1934) to contain a toxin in its epidermis. Nitta (1934) isolated a crystalline toxin, which he called *nereistoxin* and subsequently reported on its pharmacological activities. In 1960, Hashimoto and his colleagues initiated a study on the toxin's chemistry, using the annelid *L. brevicirea*, and in various papers described it as an amine with a 1,2-dithiolane ring. Its formula was $C_5H_{11}NS_2$ and the structure they proposed was:

$$
\begin{array}{c}
CH_3 \\
\diagdown \\
N-\!\!\!\begin{array}{c}\diagup\!\!\!\diagdown S \\ \phantom{x} \parallel \\ \diagdown\!\!\!\diagup S\end{array} \\
\diagup \\
CH_3
\end{array}
$$

This structure was confirmed by Konishi (1970). On the basis of Nitta's observation that flies which came in contact with the worm lost their balance and often died, and the observations of others who noted the toxicity of extracts of the worm to arthropods, nereistoxin was studied as an insecticide. A large number of related compounds were synthesized, giving rise to the production of an important agricultural insecticide. In 1972, production of this insecticide reached an estimated 1500 tons (Hashimoto, 1979).

Members of the genus *Glycera* have a proboscis equipped with four cuticular jaws, which can be everted rapidly to grasp or tear food and pull its prey back into the mouth. The jaws have curved fangs each connected by a duct with a venom gland. The gland lies within the pharynx.

The venom secreted from the venom gland of *G. convoluta* Keferstein contains a group of macromolecular toxins. Michel and Keil (1975) characterized two proteolytic activities after their separation on Sephadex G-75 and G-200. A toxin with a suggested molecular weight of 110 000–120 000 was identified after chromatography on agar-polyacrylamide gel electrophoresis. It produced a paralytic effect on the heart of *Daphnia*. A group of proteinases, having molecular weights in the 60 000–70 000 range, degraded native insoluble collagen. Another proteinase of similar molecular weight acted on the synthetic substrate of trypsin, benzoyl-L-arginine ethyl ester and was inhibited by a specific trypsin inhibitor.

*Priapulus caudatus*, a marine Ashelminthes of cylindrical shape and warty appearance, with an introversible presona has been found to eject a brown, viscous, foamable alkaline digestive fluid through its mouth. The material was found to have a protein concentration of 50 mg/ml fluid with a pH of 8·8–9·4. It had proteolytic activity with an optimum around 9·0 and substrates specific for trypsin and chymotrypsin were hydrolysed. No dipeptidase, lipase or amylase activity was observed. The toxin caused a sustained contraction of the isolated radular muscles of the whelk *Buccinum*. It was suggested that the strong proteolytic activity was associated with the animal's carnivorous feeding (Nilsson and Fänge, 1967).

# VIII. Arthropoda

The arthropods (Gr. *arthros* = joint + *podos* = foot) make up the largest phylum in the animal kingdom. There are more than 780 000 species and some of them are the most abundant individuals on earth. The arthropods include the crabs, shrimps, barnacles and other crustaceans, insects, spiders, scorpions, ticks and their allies, centipedes and millipedes. Their bodies are segmented externally in varying degrees and often show a divided head, thorax and abdomen. The appendages are joined and external surfaces are covered by an organic exoskeleton containing chitin. They have a complete digestive tract with a terminal anus, a lacunar circulatory system and respire by means of gills, air ducts, book lungs or through the body surface.

## A. *Poisoning*

Among the poisonous arthropods are certain of the crabs. The history of marine crab poisoning dates back to 1876 (Cleland, 1916). Cook Island natives were reported to eat the white-shelled crab "*Angatea*" during certain times of the year for purposes of suicide. At other times the crab was said to be non-poisonous. Rich (1882) describes a "crayfish" poisoning involving about 100 persons at Table Bay, South Africa, while numerous other reports of crab poisoning have been reported. The interested reader is referred to Halstead (1965) and Hashimoto (1979) for a more detailed accounting of marine crab poisoning. In his review on poisonous crabs, Holthuis (1968) discusses the various reasons for these arthropods being poisonous, or thought to be poisonous, and provides a list of implicated species. Additional species and a good deal of marine chemistry would need to be added to this review to bring it into present focus.

The toxicity of the sand crab *Emerita analoga* was confirmed experimentally by Sommer (1932). The toxin was found to be paralytic shellfish poison, saxitoxin or *Gonyaulax* poison. Macht *et al.* (1941) found that at certain times muscle extracts from certain lobsters and shrimp were toxic to mice. Another report too frequently overlooked is that by Banner and Stephens (1966). They used a simple but effective method for extracting a toxin from the horse-shoe crab, *Carcinoscorpinus rotundicaudata* (Latreille) and *Tachypleus gigas* (Müller) and found it toxic to mice.

Hashimoto *et al.* (1967b) screened various tissues of the crabs *Zosimus aeneus* (Linnaeus) and *Platypodia granulosa* (Ruppell) following sporadic outbreaks of crab poisoning in the Ryukyu and Amami Islands and found a poison which they assumed to be ciguateric. However, they noted that "the difference in pharmacological and chemical nature between the toxin in ciguateric fishes and in our crabs may be too great to allow us to accept the prevalent belief of the inhabitants in the Ryukyu and Amami Islands, who attribute the ciguatoxin in fish to these crabs". In the following year, Konosu *et al.* (1968) identified the toxin as saxitoxin and Mori *et al.* (1968) found that extracts of several crab species affected nerve membranes in mice and frogs.

*Atergatis floridus* has been found to be toxic (Inoue *et al.*, 1968), while the toxicity of *Eriphia sebana* seems open to question (Mote *et al.*, 1970). The toxicity of some *Demania* species was demonstrated by Garth and Alcala (1977) and by Alcala and Halstead (1970). The coconut crab, *Birgus latro* (Linnaeus) was demonstrated to be toxic to humans by Hashimoto *et al.* (1969b) and by Bagnis (1970), while Fusetani *et al.* (1980) studied 10% acetic acid extracts of the hepatopancreas and intestines of the crab and demonstrated their toxicity to mice.

Hashimoto (1979) has shown the distribution of crab toxin in the various parts of the animal (Table IX). If the lethal dose for saxitoxin to humans is used as a reference, he notes that a 0·5 g portion of the muscle of the chela of *Z. aeneus* may contain sufficient poison to kill an individual. Having reviewed approximately 30 deaths from crab poisoning found in the literature and followed in greater detail through the efforts of Professor Hashimoto, Dr. T. Sahara, Dr. H. Sato and Dr. L. M. Cummings, I would feel that this is a reasonable estimate for the one species and that the lethal dose for other toxic species is at least 3–10 times that amount.

Teh and Gardiner (1974) found that the crab, *Lophozozymus pictor*, found off the coral reefs of Singapore was toxic. An aqueous extract of the crab was

TABLE IX. DISTRIBUTION OF TOXIN IN THE BODY OF TOXIC CRABS

| Body part | Toxicity (mouse units/g) | Body part | Toxicity (mouse units/g) |
|---|---|---|---|
| Z. aeneus | | A. floridus | |
| Appendages | | Appendages | |
| Exoskeleton of chela | 2000 | Exoskeleton | 150 |
| Muscle of chela | 6000 | Muscle | 400 |
| Exoskeleton of walking legs | 2000 | Cephalothorax | |
| Muscle of walking legs | 3500 | Exoskeleton | 65 |
| Cephalothorax | | Muscle | 45 |
| Exoskeleton | 2000 | Viscera | 25 |
| Muscle | 40 | Gills | < 20 |
| Viscera | 1300 | P. granulosa | |
| Endophragm | 80 | Appendages | |
| Gills | 25 | Chelae | 400 |
| | | Walking legs | 500 |
| | | Cephalothorax | 110 |

From Hashimoto, 1979. References in original table.

subjected to a 6-stage purification process, including separation on Sephadex G-50, and found in mice to have an $LD_{50}$ of $0.377$ mg/kg, with a rate–dose relationship that differed markedly from saxitoxin and tetrodotoxin. They suggested a molecular weight of 1000–5000 for their toxin which gave reactions suggesting it contained free amino and phenol groups.

The clinical syndrome of crab poisoning is easily differentiated from that caused by bacterial poisoning and allergic shellfish poisoning. The onset of symptoms usually occurs within 3 h, with nausea, followed by severe vomiting and diarrhoea. Dizziness, drowsiness, weakness of the limbs, numbness of the lips, and various paraesthesia, aphasia, muscle paralysis, a feeling of burning in the throat and stomach and coma have all been reported and are dependent upon the crab and the amount consumed. In 530 cases reported in the literature, there have been 37 deaths.

## IX. Bryozoa (Polyzoa, Ectoprocta)

Bryozoans (Gr. *bryon* = moss + *zoon* = animal) are sessile tufted or branched cell colonies made up of individual zooides and inhabiting, for the most part, the seas and oceans, where they are usually found in shallow waters. There are approximately 4000 species of bryozoans, most of which are found attached to benthic animals or objects, such as kelp, shells or rocks. Some species are matlike, while others form incrustations. Bryozoans are often mistakenly included among the seaweeds and while some resemble colonial hydroids and corals, their internal structure is far more complicated.

## A.  *Poisoning*

Clinically, the most important bryozoan is the "curty weed" or "sea chervil," *Alcyonidium gelatinosum* Linnaeus or *A. hirsutum* (Fleming) of the family Alcyonidiidae which is responsible for Dogger Bank itch. *A. gelatinosum* is found in great numbers off the Dogger Bank (Bonnevie, 1948) and more recently it has been noted as a problem in the Baie de Seine area (Audebert and Lamoreux, 1978) and off Denmark (Carlé *et al.*, 1982). Colonies of this species are free-growing and may reach a height of more than 50 cm.

There has been considerable confusion over the past five decades as to the causative organism of Dogger Bank itch. At various times the disorder has been attributed to a diatom, plankton, sponge, cnidarian, or an alga. It now seems quite clear that the offending culprit is *A. gelatinosum*, although many references to the disorder implicate *A. hirseitum*. *Fragilaria striatula* Lyngb. has also been implicated (Fraser and Lyell, 1963). The taxonomy and, perhaps, identification of the species is not firmly established (Thorpe and Ryland, 1979).

The bryozoan is generally taken in trawling nets and while the fishermen are removing their edible catch, or while cleaning or repairing their nets, they come in direct contact with clusters of the organism. It is not universally accepted that the Dogger Bank itch is an allergic or sensitization phenomenon, although most evidence appears to indicate that this is so. Sensitization seems to vary with the extent and frequency of the exposure and, perhaps, with the individual. Continuous exposure over six months to many years is the usual history given by fishermen who have the disorder. Once sensitized, the individual may develop an allergic reaction in the presence of the bryozoan, whether it be on board ship, on the dock, or anywhere in close proximity to the organism. This would seem to indicate that the offending component of the bryozoan can be "released" from the organism and become airborne. On the other hand, there is some clinical evidence that the disorder can occur from an initial contact with the animal. However, in such cases there is an indefinite history of previous exposure and often a history of many allergies, particularly to sea foods.

The incidence of Dogger Bank itch among North Sea fishermen is not known. However, of approximately 800 fishermen who had contact with the bryozoan and who were subsequently studied, about 10% had some history of the disorder. During my visit to Lowestoft, Copenhagen and Ijmuiden during 1958, I found that among the 50 fishermen who had handled the animal over 1–10 years, six had episodes of dermatitis on contact with the organism. I would suspect that repeated handling of *A. gelatinosum* eventually leads to a high incidence of dermatitis. It is interesting to note that as far back as 1939 the Danish government passed an act, the Danish Workmen's Compensation Act, which included this disorder as an occupational hazard.

In 1966, Turk *et al.* partially purified a toxic principle by column chromatography on Sephadex G-10. Four fractions were eluted. The first had an estimated molecular weight of greater than 700 and this fraction and the third fraction proved to be active to sensitized guinea-pigs. The third fraction, having a molecular weight of approximately 250–350, was more active than the first, appeared to contain acidic groups and it was suggested that the material might have an aromatic or heterocyclic structure. Subsequently, Carlé and his group at the University of Copenhagen have carried out extensive chemical studies on the active toxic fraction of *A. gelatinosum*. The water soluble active portion was subjected to ethanol extraction, evaporation, ion-exchange chromatography, and nuclear magnetic resonance and mass spectrometry investigations which gave data consistent with the presence of a hapten, (2-hydroxyethyl)dimethyl sulphoxonium:

$$\begin{array}{c} \phantom{H_3C}\overset{\textstyle O}{\phantom{x}} \\ H_3C \diagdown \phantom{x}\Big\uparrow \\ \phantom{H_3C}S^+ - CH_2CH_2OH \\ H_3C \diagup \end{array}$$

Due to its permanent positive charge it is strongly hydrophilic. A synthetic hapten was also produced. Approximately 5 ppm of the wet weight of the bryozoan is said to be the low molecular weight hapten (Carlé *et al.*, 1982).

Pharmacological studies on the purified material have not been done. Identification of the active substance has depended upon patch testing in human subjects known to be sensitive to the bryozoan. Two patients, one suspected of having Dogger Bank itch and the other having had recurrences of the disorder over a 20- to 30-year period, displaced positive reactions with both the natural and synthetic haptens. In earlier studies (Turk *et al.*, 1966; Dubos *et al.*, 1977), sensitization of guinea-pigs was used as a biological assay. Improved recent techniques with this procedure, perhaps combined with ELISA, RIA or complement studies might be employed. Carlé *et al.* (1982) noted the formation of a precipitate on addition of the hapten to human serum. This phenomenon is being investigated.

Clinical descriptions of Dogger Bank itch have been given by Bonnevie (1948), Guldager (1959), Seville (1957), Fraser and Lyell (1963), Newhouse (1966), Audebert and Lamoureux (1978) and Carlé and Christophersen (1980). There is also a good review of the clinical problem in the discussion following the paper by Turk *et al.* (1966). Essentially, the disorder begins as a reddening and slightly irritating or itching, usually of the flexor surfaces of the forearms and volar surfaces of the hands. On repeated or continued exposure the dermatitis may spread to involve the arms, shoulders, chest and face. Severe pruritus and desquamation of the affected area may occur, and the involved parts may become quite painful to touch and to temperature changes. Angioneurotic oedema and respiratory discomfort occur in the more serious cases and joint involvement and wheezing have been reported.

# X. Echinodermata

Echinoderms are characterized in most cases by radial or meridional symmetry, a calcareous endoskeleton made up of separate plates or ossicles which often bear external spines, a well-developed coelom, a water-vascular system, and a nervous system but no special excretory system. Most echinoderms are benthic and all are marine. Approximately 85 of the 6000 species comprising the four classes (Asteroidea, Ophiuroidea, Echinoidea, Holothuroidea) are known to be venomous or poisonous. Some of the more important toxic starfishes, sea urchins, and sea cucumbers are noted in Table X.

TABLE X. SOME ECHINODERMS KNOWN TO BE VENOMOUS OR POISONOUS

| Name | Distribution |
|---|---|
| **ASTEROIDEA** | |
| Acanthasteridae | |
| *Acanthaster planci* (Linnaeus) | Widespread, Indo-Pacific, Polynesia to Red Sea, East Africa, North Australia, China, Japan, East Indies, Hawaii |
| Asteriidae | |
| *Aphelasterias japonica* (Bell) | Japan |
| *Asterias amurensis* Lütken | Japan |
| *Asterias rubens* Linnaeus | Iceland, British Isles to Senegal |
| *Marthasterias glacialis* (Linnaeus) | Eastern North Atlantic, Mediterranean to Cape Verde, Azores |
| *Patiriella calcar* (Lamarck) | Australia, Tasmania, New South Wales |
| *Pycnopodia helianthoides* (Brandt) | California to Alaska |
| Asterinidae | |
| *Asterina petinifera* Müller and Troschel | Japan |
| Astropectinidae | |
| *Astropecten scoparius* Valenciennes | Japan |
| Echinasteridae | |
| *Echinaster sepositus* Gray | English Channel to Cape Verde, Mediterranean |

Solasteridae
   *Solaster papposus* (Linnaeus)        Circumpolar, European seas south to
                                      the English Channel

OPHIUROIDEA
  Ophiocomidae
   *Ophiocomina nigra* (Abildgaard)    Norway to the Azores, Mediterranean

ECHINOIDEA
  Arbaciidae
   *Arbacia lixula* (Linnaeus)         Mediterranean, West Africa, Azores,
                                      Brazil

Diadematidae
  *Centrostephanus rodgersi* (A. Agassiz)  Australia, New Caledonia, New
                                    Zealand
  *Diadema antillarum* Philippi      West Indies, South Atlantic to
                                    Brazil, East Atlantic
  *Diadema paucispinum* (A. Agassiz)  Hawaii, South Pacific Islands
  *Diadema setosum* (Leske)        Indo-Pacific, Polynesia to East
                                    Africa, Bay of Bengal, China,
                                    Japan, Red Sea, Persian Gulf, East
                                    Indies
  *Echinothrix calamaris* (Pallas)    Indo-Pacific, Polynesia to East
                                      Africa, Bay of Bengal, Australia,
                                    Japan, Red Sea, East Indies, Hawaii
  *Echinothrix diadema* (Linnaeus)   Indo-Pacific, Polynesia to East
                                      Africa, Australia, Japan, Red Sea,
                                    East Indies, Hawaii

Echinidae
  *Echinus acutus* Lamarck        European seas, Iceland to
                                    Mediterranean
  *Paracentrotus lividus* Lamarck    Atlantic coast of Europe to West
                                    Africa, Mediterranean, Azores
  *Psammechinus microtuberculatus*  Mediterranean and Adriatic
    (Blainville)                    Seas, Azores

Echinometridae
  *Heterocentrotus mammillatus*     Indo-Pacific, Polynesia to
    (Linnaeus)                    East Africa, Hawaii, Australia, Red
                                      Sea, East Indies, China, Japan

Echinothuridae
  *Araeosoma thetidis* (Clark)      East coast of Australia, North
                                    Island of New Zealand
  *Asthenosoma varium* Grube     Indonesia, Indian Ocean, Suez (Red
                                    Sea), East Indies, China, Japan
  *Phormosoma bursarium* Agassiz   Pacific, Indo-Pacific, Natal,
                                    (Hawaii to Arabian Sea)

Spatangidae
  *Spatangus purpureus* Müller     Northern Europe to West Africa,
                                    Mediterranean Sea

TABLE X. (CONTINUED)

| Name | Distribution |
|------|--------------|
| Strongylocentrotidae | |
| *Strongylocentrotus drobachiensis* (Müller) | Circumpolar south to Scotland, Chesapeake Bay, Puget Sound |
| *Strongylocentrotus purpuratus* (Stimpson) | Vancouver to Lower California (North American West Coast) |
| Toxopneustidae | |
| *Lytechinus variegatus* (Lamarck) | West Indies, North Carolina, South Atlantic to Brazil |
| *Sphaerechinus granularis* (Lamarck) | Mediterranean, Channel Islands to Cape Verde, Azores |
| *Toxopneustes elegans* Döderlein | Japan, Ryukyu Islands |
| *Toxopneustes pileolus* (Lamarck) | Indo-Pacific, Melanesia to East Africa, Japan, East Indies, Bay of Bengal |
| *Tripneustes gratilla* (Linnaeus) | Indo-Pacific, Polynesia to East Africa, Red Sea, Australia, Japan, Bay of Bengal, East Indies, Hawaii |
| *Tripneustes ventricosus* (Lamarck) | West Indies south to Brazil, West Africa |

HOLOTHUROIDEA

| | |
|------|--------------|
| Chirodotidae | |
| *Polycheira rufescens* (Brandt) | Japan, East Africa, East Indies, Bay of Bengal, Australia, Philippines, South Pacific |
| *Stichopus chloronotus* Brandt | Indo-Pacific, Australia, East Africa, Bay of Bengal, East Indies, Japan, Hawaii |
| *Stichopus japonicus* Selenka | Japan |
| *Stichopus variegatus* Semper | Polynesia to Indian Ocean, Australia, Red Sea, East Africa, East Indies, Japan |
| Cucumariidae | |
| *Cucumaria echinata* von Marenzeller | Japan |
| *Cucumaria japonica* Semper | South Pacific, East Indies, Japan |
| *Neothyone gibbosa* Deichmann | Gulf of California, Peru |
| *Pentacta australis* (Ludwig) | Tropical West Pacific, East Indies, Australia |
| Holothuriidae | |
| *Actinopyga agassizi* (Selenka) | West Indies |
| *Actinopyga lecanora* (Jaeger) | Indian Ocean, Australia, Celebes, East Africa, Micronesia, Ryukyu Islands, East Indies |
| *Holothuria* (*Microthele*) *nobilis* (Selenka) | Australia, East Africa, Red Sea, Ceylon, East Indies, China, Japan, Philippines, Hawaii, South Pacific |

| | |
|---|---|
| *Bohadschia argus* Jaeger | Indo-Pacific, Polynesia, Ryukyu Islands, Bay of Bengal, East Indies, Australia |
| *Bohadschia bivittata* (Mitsukuri) | Philippines, Japan, China, South Pacific, Ryukyu Islands |
| *Holothuria* (*Halodeima*) *atra* Jaeger | Indo-Pacific, Polynesia, Australia, East Africa, Red Sea, Bay of Bengal, East Indies, Japan, Hawaii |
| *Holothuria axiologa* H. L. Clark | Australia, Palau |
| *Holothuria forskåli* | |
| *Holothuria* (*Semperothuria*) *imitans* Ludwig | South Pacific, Ceylon |
| *Holothuria* (*Brandtothuria*) *impatiens* (Forskål) | Circumtropical |
| *Holothuria* (*Ludwigothuria*) *kefersteini* (Selenka) | Mexico to Peru and offshore islands |
| *Holothuria* (*Mertensiothuria*) *leucospilota* Brandt | Australia, East Africa, Red Sea, Bay of Bengal, East Indies, Philippines, South Pacific, Hawaii, Japan, China |
| *Holothuria* (*Selenkothuria*) *lubrica* Selenka | Gulf of California to Ecuador and Galapagos |
| *Holothuria monacaria* (Lesson) | Fiji and Samoa, New Guinea, Timor, Queensland (Indo-Pacific) |
| *Holothuria poli* Delle Chiaje | Mediterranean and adjacent Atlantic coasts |
| *Holothuria* (*Metriatyla*) *scabra* Jaeger | East Africa, Red Sea, Bay of Bengal, East Indies, Australia, Japan, South Pacific, Philippines, Indian Ocean |
| *Holothuria tubulosa* Gmelin | Mediterranean, adjacent Atlantic coast |
| *Holothuria vagabunda* Selenka | Tropical western Pacific |
| *Patinapta ooplax* (von Marenzeller) | East Africa, Japan, East Indies, New Guinea |

Molpadiidae

| | |
|---|---|
| *Paracaudina australis* (Semper) | Australia |
| *Paracaudina chilensis* (J. Müller) | Japan, Australia, Bay of Bengal |
| *Thelenota ananas* (Jaeger) | Indo-Pacific, Polynesia, Micronesia, Australia |

Phyllophoridae

| | |
|---|---|
| *Afrocucumis africana* (Semper) | Japan, East Africa, Bay of Bengal, East Indies, Australia |
| *Euthyonidium* sp. Deichmann | West Indies and American West Coast |

Stichopodidae

| | |
|---|---|
| *Parastichopus nigripunctatus* (Augustin) | Japan |

Synaptidae

| | |
|---|---|
| *Euapta lappa* (Müller) | West Indies |

Asteroids, starfishes, or sea stars, have a central disc and five or more tapering rays or arms. On the upper surface are many thorny spines of calcium carbonate in the form of calcite intermingled with organic materials. Between the spines are small, soft papulae that project from the body cavity and serve as gills and for excretion. Also around the spines are minute *pedicellariae* which serve to keep the surface free of debris and aid in the capture of food. The largest sea stars have spreads of approximately 80 cm.

The regular sea urchins have rounded radially symmetrical bodies which are enclosed in a hard calcite shell from which calcareous spines and pedicellariae arise. The spines may be straight and pointed, curved, flat-topped, club-shaped, oar-shaped, umbrella-shaped, thorny, fan-shaped or hooked. They may vary in length from less than 1 mm to over 30 cm. The spines serve in locomotion, protection, digging, feeding, and producing currents; certain of the primary and secondary spines bear poison glands.

The sea cucumbers are soft-bodied animals covered by a leathery skin that contains only microscopic calcareous plates. The body is elongated in the oral–aboral axis. The mouth is surrounded by 10–30 tentacles, comparable with the oral tube feet of other echinoderms. The coelom is large and fluid filled. Movement of the muscles over the fluid-filled body enables the animal to extend or contract its body and to carry out wormlike movements. The largest sea cucumber, *Synapta maculata*, may measure 2 m in length when less than 5 cm in diameter. According to Nigrelli and Jakowska (1960) at least 30 species belonging to four of the five orders of Holothuroidea are toxic. Some toxic species, *Thelenota ananas* (Jaeger), *Stichopus variegatus* (Semper), *Holothuria atra* (Jaeger) and *H. axiologa* (H. L. Clark), are esteemed as food in the Orient.

### A.   Venom or Poison Apparatus

In the asteroids the calcite spines are covered by a thin integument composed of an epidermis and a dermis. Within the epidermis are two kinds of gland cells, an acidophilic cell and a basophilic cell. It is believed that the former cells produce the toxin while the latter are involved in mucus production. The toxin is discharged into the water or, as in the case of humans, directly onto the skin. In addition, sea stars have pedicellariae, which contain poison glands in the concave cavity of their valves. These are described in the next section. Some sea stars produce poisoning following their ingestion.

The principal venom apparatus in the sea urchin, heart urchin and sand dollar is the pedicellaria. In essence, pedicellariae are modified spines with flexible heads. There are four primary kinds: ophiocephalous, dentate, globiferous and foliate (Fig. 11). Some urchins possess all four kinds;

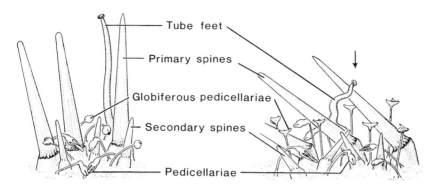

FIG. 11. Diagrammatic sketch of relationship between pedecellariae, tube feet and spines in a sea urchin. ↓ indicates stimulus.

however, many echinoids do not. The pedicellariae function is food getting, grooming and self-defence. The glandular, gemmiform, or globiferous type pedicellaria serves as a venom organ. In most echinoids the so-called "head" of the pedicellaria is composed of three calcareous jaws or valves, each having a rounded, tooth-like fang. The jaws are usually invested in a globose, fleshy and somewhat muscular sac which possesses a single or double gland over each valve (Fig. 12).

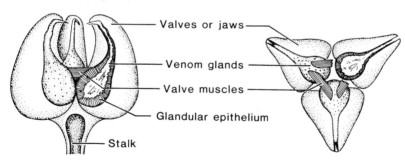

FIG. 12. Head of pedicellaria showing the three calcareous jaws with their venom glands.

A second trilobed gland system, anatomically and histologically distinct from the head gland is present in some urchins. These are located in the thin stem or stalk of some globiferous pedicellariae and certain specialized ophiocephalous forms. These various structures and the mechanisms for ejecting the venom have been described in more detail by Hyman (1955), Nicols (1962), Halstead (1965), and the work edited by Boolootian (1966). The venom is formed in these head or stalk glands and discharged following certain kinds of stimulation.

O'Connell (1971) studied the fine structure of the globiferous pedicellariae from the purple sea urchin, *Strongylocentrotus purpuratus* (Stimpson). He found secretory cells lining the inside of the venom gland. They were typically mucoid in appearance and possessed small volumes of basally displaced, vesiculated cytoplasm and an extensive system of vacuoles which dominated the apical nine-tenths of each cell. The vacuoles contained ground substances of various densities. The venom-producing cells contained numerous cytomembrane complexes, indicating metabolic activity. The vacuolar products were tentatively identified as dense basophilic fuchsin-positive spherules, apparently formed of protein and less dense, granular, alcianphilic compounds containing nonsulphated acid mucosubstances. O'Connell concluded that the gland cells were probably holocrine in function and release their vacuolar contents upon constriction of the muscles surrounding the gland sac.

The primary and secondary spines of some urchins have specialized organs containing a gland, which is said to empty its contents through the hollow-spine tip under certain conditions. We found a toxin in the secondary spines of *Echinothrix calamaris* and *E. diadema* but not in their primary spines. The secondary spines of *Asthenosoma varium* Grube and *Araeosoma thetidis* (H. L. Clark) and the primary spines on the oral side of *Phormosoma bursurium* also contain a venom (Mortensen, 1929; Hyman, 1940). According to Halstead (1978) the spine venom glands are best developed in the secondary aboral spines of *A. varium*.

In the class *Holothuroidea*, the sea cucumbers, some members possess special defence organs known as Cuvierian tubules, which arise from a common stem of the respiratory tree. When these animals are irritated they emit these organs through the anus. The tubules become elongated by hydrostatic pressure so that once through the anus they become extremely sticky threads in which the attacking animal becomes ensnared. As noted by Endean (1957), the process of elongation may split the outer layer of covering cells, thereby releasing a proteinaceous material which forms an amorphous mass having strong adhesive properties. These elongated threads may attain a length of almost a metre. They separate from their attachment and are left behind as the holothurian crawls away. The autotomized parts are regenerated in time.

In some sea cucumbers, however, as in *Actinopyga agassizi* Selenka, the tubules do not become sticky, nor do they elongate, but they are eviscerated in a somewhat similar manner and they discharge a toxin from certain highly developed structures filled with granules. The toxin is capable of killing fishes and other animals. Evisceration may be provoked by handling the animal, either in or out of the water, and by excessive changes in temperature, pH and oxygen balance. In other sea cucumbers, as in *Holothuria atra*, which do not possess Cuvierian tubules, the toxin may be discharged through the body wall. In Guam, natives cut up the common black sea cucumbers and squeeze the contents of the animal into crevices and pools to deactivate fish. "Before long the water became noticeably turbid in appearance, and shortly thereafter fish began coming to the surface of the pool, exhibiting much the same type of behaviour as in rotenone poisoning" (Frey, 1951). I have seen the same technique used in Saipan, where the fishermen wear goggles to protect their eyes from the irritating substances of these animals.

## B. *Chemistry and Toxicology*

Many echinoderms secrete a mucus or liquid from their integument that appears to play a role in their defensive armament. The chemical nature of these substances varies considerably as, obviously, does their pharmacological activity. For instance, the viscous discharge from the massive multicellular integumentary glands of the brittle star, *Ophiocomina nigra* (Abildgard) is characterized as a highly sulphated acid mucopolysaccharide, containing amino sugars, sulphate esters and other substances complexed to proteins (Fontaine, 1964). The pH of this discharge is approximately 1, which probably makes it very offensive to other marine animals which might seek to prey upon it.

Among other substances isolated from the echinoderms is a quaternary ammonium base ($C_7H_7NO_2$, picolinic acid methyl betaine), known as homarine (Hoppe-Seyler, 1933; Gasteiger *et al.*, 1960), several phosphagens (phosphoarginine and phosphocreatine), sterols (Fagerlund and Idler, 1959; Bergmann and Domisky, 1960; Scheuer, 1973), saponins (Hashimoto and Yasumoto, 1960; Yasumoto *et al.*, 1966; Ruggieri *et al.*, 1970), and other compounds (Hashimoto, 1979).

Perhaps, one of the difficulties, again, in determining the toxic substance(s) in the echinoderms is related to the differences in the origin of the test material, that is, not only the chemical variations in different species but the differences in the kind of product extracted from the animals. In the sea star, for example, some investigators grind up the fresh dead asteroid, others use frozen specimens, while still others have dried the animal and prepared extracts, or used the dried meal of the starfish. Finally, some investigators use the material elaborated from the tube feet after stimulation, either as a fresh product or dried or lyophilized, or that obtained from the animal's discharge in water. To relate these various products chemically is not easy. Pharmacologically, the task is more difficult, not only because of the differences in products but the differences in the bioassays that have been employed. Some investigators use the toxicity to fish assay, while others use a haemolytic test, an oral or parenteral mammalian assay, and still other researchers use an intact mollusc or other invertebrate, measuring the withdrawal response or lethality. Finally, a few investigators use specific marine isolated tissue preparations.

The chief problem in relating the data from these various studies is their significance to each other, to what we mean by toxicity and, most importantly, to the function of that toxicity in the animal and its environment. For instance, what is the relationship between haemolysis and tube-foot withdrawal? Is the same substance involved? Does parenteral injection give us the ecological significance of an ingested toxin? Hashimoto (1979) has done an excellent job of reviewing the chemistry and pharmacology of echinoderm toxins and he has been wise enough to avoid attempting to interpret unrelated and sometimes confusing data. This author follows suit. The time is not yet right for any definitive attempt to put the data into meaningful categories. However, I will attempt to present some generalized impressions, some of which carry over from what we know from more detailed studies on other marine invertebrate toxins.

It is known that sea stars liberate a bioactive substance from the epidermis of their tube feet (Feder, 1963) and that certain molluscs, and to a lesser extent some anemones and echinoderms, respond to this material (Feder, 1959; Margolin, 1964; Montgomery, 1967). The active substance is thought to be sensed by chemoreception or by direct contact with the asteroid (Feder and Arvidsson, 1967; Montgomery, 1967) and appears to play a role in the animal's defensive posture as a repellent, and under certain concentrations and conditions it can be lethal to some fishes and marine invertebrates. The secretion also has certain other biological activities, including one related to antimicrobial function (Ruggieri and Nigrelli, 1973). Although the exact role of this function in the asteroid's armament has not been clearly established, it probably serves to protect the animal in a very changing environment, where it is widely exposed to micro-organisms or it may repel micro-organisms in the biota. In addition to this toxin there are at least several other substances in the body of the starfish that have an effect on living cells, tissues or intact animals (Nigrelli and Zahl, 1952; Shimada, 1969).

The toxin released into the environment by the starfish is now known to be a saponin or saponin-like substance (Hashimoto and Yasumoto, 1960) having a varying sugar composition but always bearing a sulphate. This toxin is a potent surfactant and causes fish and marine invertebrates, in general, to retreat, to attain a defensive posture, or to react in avoidance. Rio and associates (1963) extracted a saponin-like substance from the sunburst starfish, *Petasometra helianthiodes* A. H. Clark, which was similar to holothurin, although they differed in their sugar moieties.

Yasumoto *et al.* (1964) isolated a substance from the dried meal of *Asterina petinifera* that was toxic to fish and was haemolytic. They concluded that the toxic material was a saponin. Using specimens of fresh *Asterias amurensis* they found that their partially purified product was haemolytic, toxic to killifish, lethal to fly maggots and earthworms, and causes vomiting in cats

(Yasumoto *et al.*, 1964). On thin layer chromatography the toxin was resolved into six components (Yasumoto and Hashimoto, 1965). The major component was named asterosaponin A and the second major component was called asterosaponin B. Their chemical properties are shown in Table XI. Using various chromatographies, Ikegami *et al.* (1972) obtained two bioactive fractions from sea stars which were similar to asterosaponins A and B.

TABLE XI. PROPERTIES OF ASTEROSAPONINS A AND B

| Property | Asterosaponin A | Asterosaponin B |
|---|---|---|
| Melting point | 185–190°C | 189–191°C |
| Specific rotation $[\alpha]_D^{20}$ | $+ 0.03°$ | $- 2.66°$ |
| Haemolytic index | 55 600 | 41 000 |
| Molecular formula | $C_{50-52} H_{88-90} O_{25-26}$ SNa | $C_{56} H_{101} O_{32}$ SNa |
| U.V. absorption | 244 nm | 248 nm |
| I.R. absorption | 1640 cm$^{-1}$ | 1700, 1640 cm$^{-1}$ |
| Sugar | D-Quinovose, D-Fucose (2:2) | D-Quinovose, D-Fucose, D-Xylose, D-Galactose (2:1:1:1) |
| Sulphate group | One mole | One mole |
| Aglycone | Asterogenin-I, II | Asterogenin-I, II |

From Hashimoto, 1979.

Feder and Arvidsson (1967) demonstrated that partially purified drippings from thawed *Asterias rubens* and *Marthasterias glacialis* caused marked avoidance reactions in *Buccinum* and several species of ophiuroids, and strong muscle contractions in *Buccinum undatum* radula muscle preparations. They speculated that since the reactive substances are produced in the same external epithelium of the oral and aboral surfaces of sea stars, that possibly all predatory sea stars of the order Forcipulata might exude similar substances causing avoidance reactions.

Mackie and his group at the University of Aberdeen (1968, 1970) isolated and partially characterized a surface-active saponin from *A. rubens*. It had an approximate molecular weight of 514 and paper chromatography revealed that this major steroid was present with at least four other steroid components. The major saponin, saponin 1, was found to provoke a large part, if not all, of the activity of the extract toward the intact snail response. Its chemical characteristics seem similar or identical to the asterosaponin described by Yasumoto and Hashimoto (1965). The active principle in *Marthasterias glacialis* also behaved in a similar manner chemically, and its haemolytic and foot-muscle bioassay effects were similar. A second saponin, saponin 2, had different characteristics. Mackie and colleagues concluded that the saponins or saponin-like substances were surface active, eliciting either an immediate avoidance reaction or, as in the mollusc *Buccinum undatum*, a series of convulsions. They developed a sensitive, semi-quantitative assay, using the withdrawal response of the intact mollusc's foot-muscle. Positive correlations between the intact preparation and the tube-foot were demonstrated. A further bioassay, the isolated radular muscle of *Buccinum*, showed that in order to elicit a significant response the presence of a protein fraction was essential.

Owellen *et al.* (1973) isolated a "lethal" saponin fraction from *A. vulgaris* which contained four major components. It had cytolytic activity against mouse, rat and human erythrocytes and lymphocytes. Although an $LD_{50}$ determination was not made, mice survived intraperitoneal doses of 50 mg/kg body weight but succumbed to 200 mg/kg. Croft and Howden (1974) isolated and partially characterized a haemolytic steroidal saponin from the sea star *Patiriella calcar*, which they called calcarsaponin. Its aglycone, calcargenin, was partially characterized and had a molecular weight of 414. The sugars in calcarsaponin were galactose, fucose and quinovose, and the saponin contained a sulphate moiety. Fleming *et al.* (1976) found a mixture of four steroidal saponins in *Acanthaster planci*. The major sapogenins were 5α-pregn-9(11)-ene-3β,6α-diol-20-one and 5α-cholesta-9(11), 20(22)-diene-3β,6α-diol-23-one, previously identified. Spectral evidence was obtained bearing on the structures of the remaining sapogenins. Hydrolysis of the saponins also yielded quinovose, fucose, arabinose, xylose, galactose, glucose and sulphate. The highly surface-acting properties of the asteroid saponins make them haemolytic, anticoagulant, cytolytic and toxic to fishes. They also have antimicrobial properties, inhibit cancer cells, and induce abnormalities during the development of fertilized sea urchin eggs.

The first observation that certain pedicellariae were venomous was probably that of Von Uexkull (1899), who demonstrated that when the pedicellariae of several species were allowed to sting various animals, signs of poisoning developed. Henri and Kayalof (1906) found that aqueous extracts of several urchins were lethal to certain gastropods, cephalopods, crustaceans, fishes, a lizard, and rabbits but not to several echinoids or a frog. Levy (1925) demonstrated the haemolytic effect of several sea urchins pedicellarial extracts, and Fujiwara (1935) observed that when the pedicellariae of *Toxopneustes pileolus* were allowed to sting the shaved abdomen of a mouse the animal developed respiratory distress and exhibited a decrease in body temperature. Pérès (1950) found that the injection of thermostable extracts from the macerated pedicellariae of *Sphaerechinus granularis*, and certain other species, were lethal to isopods, crabs, octopods, sea stars, lizards and rabbits, although they had little effect on frogs. He thought that the toxic factor could be found in other tissues of the animal, although he felt much more was concentrated in the pedicellariae. Mendes (1963) reported the presence of a dialysable acetylcholine-like substance in the pedicellariae of *Lytechinus variegatus*. The toxin was partially deactivated by heating and completely destroyed by one hour's exposure to NaOH.

The protein nature of pedicellarial toxin was first described by Alender *et al.* (1965), who found that the active principle in *Tripneustes gratilla* (Linnaeus) was non-dialysable, thermolabile, pH stable, almost completely soluble in distilled water and precipitable with a relatively high potency in the presence of 2/3 saturated ammonium sulphate in a yield of 26% dry weight of the starting material. Sixty per cent of the lethal activity appeared to be in one fraction which had a sedimentation coefficient of 2·6 Svedberg units at 20°C. The intravenous $LD_{50}$ in mice, based on the quantity of precipitable protein nitrogen, was $1·59 \times 10^{-2}$ mg/kg body weight for the crude material and $1·16 \times 10^{-2}$ mg/kg for the protein.

The toxic protein possessed haemolytic activity against human type A and B, rabbit, guinea-pig, beef, sheep and fish erythrocytes. Intravenously, it produced a dose related hypotension that was responsive to adrenalin. It had a deleterious effect on the isolated heart and guinea-pig ileum. In both these preparations the toxin caused the release of histamine and serotonin. It seems to have little effect on isolated toad nerve (Alender and Russell, 1966). However, Parnas and Russell (1967), using the deep extension abdominal muscle of the crayfish, showed that the toxin produced a rapid block in the response of the indirectly stimulated muscle. Even with low concentrations there was an irreversible block in the muscle's response to intracellular stimulation. The compound action potential of the crayfish limb nerve was also blocked by the toxin but this potential reappeared on washing. The toxin caused considerable damage to the muscle fibres. These findings seemed to indicate that pedicellariae toxin blocks the response from both nerve and muscle and was cytolytic. A similar conclusion was gleaned from studies on the nerve-muscle preparation of the guinea-pig.

Feigen *et al.* (1966) observed that pedicellarial toxin from *T. gratilla* elicited prolonged contractions of isolated guinea-pig ileum. Chemical evidence was obtained for the release of histamine from ileal, cardiac and pulmonary tissues, as well as from the colonic and pulmonary tissues of the rat. The histamine release was quantitatively dependent on the concentration of the toxin acting upon the tissue. Subsequently, Feigen *et al.* (1968) showed that the toxin acts kinetically as an enzyme and that one of the substrates is electrophoretically pure human $\alpha_2$-macroglobulin. They also demonstrated that the crude toxin was kininolytic, with respect to the dialysable reaction-product, as well as to synthetic bradykinin. Immunoelectrophoretic analysis revealed the existence of two distinct antigenic determinants, and additional serological tests showed that antibodies directed against the pedicellarial toxin of *T. gratilla* cross-reacted with the test proteins of *Strongylocentrotus purpuratus*.

The active material obtained from the reaction between crude sea urchin toxin and heated plasma was a mixture of pharmacologically active peptides, one of which was bradykinin. The peptides could be separated by gel filtration on Sephadex and resolved into seven distinct components by paper chromatography, three of the slower moving peptides being chromatographically and pharmacologically identical with commercial bradykinin.

Fleming and Howden (1974) obtained a partly purified toxin from the pedicellariae of *T. gratilla* by extracting with two volumes of cold 0·01 M Tris-HCl buffer, pH 8·0, homogenizing the product, centrifuging, decanting the supernatant and filtering through a Millepore filter. Ion-exchange chromatography on carboxymethyl cellulose and electrofocusing showed that the toxin was acidic in nature. On electrofocusing the major components were eluted at 4·1, 5·1 and 5·4. The toxin, as established by intraperitoneal injection in mice, was at the 5·0–5·1 isoelectric point. When this fraction was chromatographed in Sephadex G-200 a molecular weight of 78 000 ± 8000 was found. This seems consistent with the sediment coefficient of 4·7 (67 000) presented by Feigen *et al.* (1968).

Apparently unaware of the recent works of Feigen *et al.* (1974), Kimura *et al.* (1975) also found that the pedicellarial venom of *Toxopneustes pileolus* was a mixture of several basic peptides having kinin-like activity. Following extraction and separating procedures they found that their urchi-toxin $F_2$ exhibited absorption maxima at 260, 280 and 330 mu. Ultracentrifugation yielded two peaks having sediment constants of 2·1 and 4·7, respectively. The approximate molecular weights were 30–40 000 and 70–80 000. Thin-layer chromatography indicated a single spot but with some trailing. The toxin was basic in character. It produced a decrease in frog heart activity and an irregular rhythm and it caused transient vasoconstriction, increased motility of rabbit ileum, contraction of guinea-pig ileum and rat uterus, and increased peripheral blood vessel permeability. The interacting activity was destroyed by chymotrypsin. On exposure of urchi-toxin $F_2$ to Sephadex G-25, three fractions ($F_3$, $F_4$, $F_5$) were obtained. Only $F_3$ had significant biological activity. $F_3$ had a lesser cardiac effect than $F_2$, while $F_4$ and $F_5$ had no effect. $F_3$ and $F_5$ produced contraction of the isolated guinea-pig ileum. When these various properties were compared with those produced by bradykinin, the conclusion was that $F_2$ had 50% bradykinin activity, while $F_3$ had 75%, $F_4$ had 0% and $F_5$ had 30%.

Cooper (1880) was apparently the first to note that some sea cucumbers can discharge tenacious filaments from their visceral cavity and that these may produce painful wounds in humans. Saville-Kent (1893) observed that eating the sea cucumber *Stichopus variegatus* can cause death, although

Cleland (1913) did not think so, and Clark (1921) wrote that the animal's Cuvierian organ was quite harmless. According to Halstead (1965), the initial studies of the poisonousness of sea cucumbers were carried out by Yamanouchi (1929) who observed that when fishes were placed in the aquarium to which aqueous extracts of *Holothuria vagabunda* were added the fish died. Subsequently, he obtained a toxic crystalline product, termed holothurin, and found it present in 24 of the 27 species of sea cucumbers examined (Yamanouchi, 1942, 1955). More recently, Bakus and Green (1974) have found that the more tropical the locality, the greater the probability that the holothurin will be toxic to fishes.

Nigrelli (1952) found that the toxic substance (holothurin) extracted from the Bahamian sea cucumber *Actinopyga agassizi* was composed of 60% glycosides and pigments, 30% salts, polypeptides and free amino acids, 5–10% insoluble protein and 1% cholesterol. The cholesterol-precipitated fraction, known as holothurin A, represented 60% of the crude holothurin and was given the empirical formula $C_{50-52}H_{81-85}O_{25-26}SNa$ (Chanley *et al.*, 1960) and I.R. absorption at 1748 and 1629 cm$^{-1}$. Holothurin appeared to consist of at least four steroid aglycones bound individually to four molecules of monosaccharides; it showed no absorption in the ultraviolet region. It was probably a mixture of aglycone of approximately 26–28 carbon and 4–5 oxygen atoms, one molecule each of four different sugars and one molecule of sulphuric acid as a sodium salt (Chanley *et al.*, 1960). A provisional structure was published by these workers but was changed by Friess and Durant (1965); see Fig. 13. It resembled digitonin and other saponins in

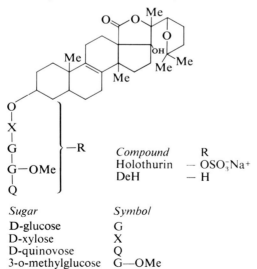

| Compound | R |
|----------|---|
| Holothurin | $- OSO_3^-Na^+$ |
| DeH | $- H$ |

| Sugar | Symbol |
|-------|--------|
| D-glucose | G |
| D-xylose | X |
| D-quinovose | Q |
| 3-o-methylglucose | G—OMe |

FIG. 13. Provisional structure for holothurin A (Friess *et al.*, 1967).

both its chemical and biological activities. The toxin had a deleterious effect on some sharks and was suggested as a shark repellent.

In 10 ppm, holothurin was found to be lethal to *Hydra*, the mollusc *Planorbis* and the annelid *Tubifex tubifex*. It has slightly greater haemolytic action than saponin and stimulated haemopoiesis in the bone marrow of winterized frogs. It also appears to have some antimetabolic activity (Nigrelli and Jakowska, 1960). Administration of $C^{14}$-labelled mevalonic acid into *S. badionotus* was incorporated into holothurin, which might indicate that the animal synthesizes holothurin *in vivo* (Murthy and Der Marderosian, 1973).

In the mammalian phrenic nerve-diaphragm preparation, holothurin A produced a contracture of the muscle, followed by some relaxation, and a gradual decrease in the recorded amplitude of both the directly- and indirectly-elicited contractions, the latter decreasing at a slightly greater rate than the former. The intravenous $LD_{50}$ in mice was approximately 9 mg/kg body weight (Friess *et al.*, 1960). In experiments in frogs, Thron and his colleagues (1963) demonstrated that holothurin A produced an irreversible block and destruction of excitability on the single node of Ranvier in the sciatic nerve. The toxin did not produce any observable damage to the axonal walls or sheath. The poison did not exert a blocking action on the *in vitro* AChE-ACh system.

Matsuno and Yamanouchi (1961) isolated the major aglycone from the holothurin A of *H. vagabunda* and named it holothurigenin. Holothurigenin, $C_{30}H_{44}O_5$, melting point 301°C, contained three hydroxyls, a five-membered lactone, and a heteroannular diene. On hydrolysis, Chanley *et al.* (1966) isolated two genins by repeated alumina column chromatography of a mixture of aglycones derived from the holothurin A of *A. agassizi*. The two genins account for 30% of the total aglycone, and were identified by spectral and chemical techniques as 22,25-oxidoholothurinogenin and 17-desoxy-22,25-oxidoholothurinogen.

Yasumoto *et al.* (1967) isolated two saponins from the body walls of *H. vagabunda* and *H. lubrica* by column chromatography on silic acid. The major saponin was obtained as needles, m.p. 225–226°C (dec.) and had an empirical formula of $C_{55}H_{99}O_{29}SNa$. Its sugar composition and other properties were identical with those of holothurin A isolated by Chanley *et al.* (1966). The other saponin, named holothurin B, was isolated as needles, m.p. 213–216°C (dec.). It had an empirical formula of $C_{45}H_{75}O_{20}SNa$ and I.R. bands at 1745 and 1640 cm$^{-1}$. On acid hydrolysis it was found to be a mixture of aglycones, sulphuric acid, and one mole each of D-quinovose and D-xylose. Matsuno and Iba (1966) also isolated holothurin B from *H. vagabunda*. Table XII shows the chemical properties of holothurins A and B as published by Hashimoto (1979). Shimada (1969) isolated a toxin from

TABLE XII. PROPERTIES OF HOLOTHURINS A AND B

| Property | Holothurin A | Holothurin B |
|---|---|---|
| Melting point | 224–226°C (dec.) | 213–236°C (dec.) |
| Specific rotation $[\alpha]_D^{20}$ | $-14{\cdot}4°$ | $-7{\cdot}3°$ |
| Haemolytic index | 187 000 | 125 000 |
| Empirical formula | $C_{55} H_{99} O_{29} SNa$ | $C_{45} H_{75} O_{20} SNa$ |
| U.V. absorption | None | None |
| I.R. absorption | 1745, 1640 cm$^{-1}$ | 1745, 1640 cm$^{-1}$ |
| Sulphate group | One mole | One mole |
| Sugar | 3-Q-Methyl-D-glucose, D-Quinovose, D-Xylose, D-Glucose (1:1:1:1) | D-Quinovose, D-Xylose (1:1) |
| Ability to form adduct with cholesterol | + | + |
| Aglycone | Mixture | Genin-I, II, III |

From Hashimoto, 1979.

the body wall of *Stichopus japonicus* which closely resembled the holothurins. He called the toxin holotoxin. It is a steroid glycoside which differs from holothurin in that it is not a sulphate and has unique antifungal properties. It is similar to holothurin in its infrared spectrum, showing bands at 1745 and 1640 cm$^{-1}$, indicative of a five-membered ring lactone and one double

TABLE XIII. OCCURRENCE OF TOXINS IN HOLOTHURIOIDEAE spp.

| Species | Substance No. | | | | | | | | | | |
|---|---|---|---|---|---|---|---|---|---|---|---|
| | 1 | 2 | 3 | 4 | 5 | 6 | 7 | 8 | 9 | 10–17 | 18/19 |
| *Actinopyga agassizi* | 0 | 0 | 0 | 0 | — | — | — | — | — | 0 | — |
| *Halodeima grisea* | 0 | — | + | — | — | — | — | — | — | — | — |
| *Holothuria tubulosa* | + | + | — | — | — | — | — | — | — | — | — |
| *Holothuria forskåli* | 0 | 0 | | | | | | | | | |
| | + | + | — | — | — | — | — | 0 | — | — | — |
| *Holothuria poli* | + | + | + | — | + | — | — | + | + | — | — |
| *Bohadschia koellikeri* | — | — | — | — | — | + | + | + | + | — | — |
| *Stichopus japonicus* | — | — | — | — | — | — | — | — | — | — | + |

+, isolated from body wall; 0, isolated from Cuvierian organs.
From Habermehl and Volkwein, 1971. References in original table.

bond, respectively. Although it has marked antifungal activity, it has little effect on Gram-positive and Gram-negative bacteria and mycobacteria.

Habermehl and Volkwein (1968, 1970, 1971) investigated the toxins from the Cuvierian organ and body wall of a number of species and summarized the findings in a table (Table XIII). According to the authors, these compounds are glycosides of tetracyclic triterpenes which are derivatives of lanosterol and were the first glycoside triterpenes derived from animals. The authors presented some representative formulas:

In their 1971 paper, Habermehl and Volkwein proposed a new nomenclature for the toxins in an attempt to bring order to the classification difficulties presented by gimmick or trivial names. Their suggestions are shown in Table XIV. Unfortunately, again, gimmick names seem to have persisted. Toxinology awaits its Linnaeus.

Working with Shimada's holotoxin, Kitagawa et al. (1974, 1976) found it was composed of three saponins, which they named holotoxin A, B and C. All three substances had antimicrobial activity and on hydrolysis yield two genins, D-xylose, D-quinovose, 3-O-methyl-D-glucose and D-glucose. Their empirical formulas are $C_{57}H_{94}O_{27} \cdot H_2O$ (holotoxin A) and $C_{65}H_{104}O_{33} \cdot H_2O$ (holotoxin B). Other substances isolated from the holothurins include stichoposides (Anisimov et al., 1973), cucumarioside C, $C_{30}H_{44}O_5$ (Elyakov and Peretolchin, 1970), and thelothurins A and B (Kelecom et al., 1976). Anisimov et al. (1974) demonstrated that cucumarioside C, extracted from the sea urchin Cucumaria fraudatrix, produced changes in sea urchin embryo development leading to alterations in biopolymer syntheses. It differed from the stichoposides in that it affected sea urchin eggs at various stages of development, whereas stichoposides uniformly arrest egg division.

TABLE XIV. NOMENCLATURE PROPOSED BY HABERMEHL AND VOLKWEIN (1971) BASED ON THE BASIC COMPOUND HOLOSTANOL
($3\beta$-$20\alpha_F$-DIHYDROXY-$5\alpha$-LANOSTAN-18-CARBOXYLIC ACID-$(18\rightarrow20)$-LACTONE)

| Compound No. | Nomenclature Basic substance: Holothurinogenin | Trivial name | Nomenclature Basic substance: Holostanol |
|---|---|---|---|
| 1 | 22,25-Epoxy-holothurinogenin | — | 22-25-Epoxy-$\Delta^{7,9(11)}$-holostadien-$3\beta$-$17\alpha$-diol |
| 2 | 22,25-Epoxy-17-desoxy-holothurinogenin | — | 22,25-Epoxy-$\Delta^{7,9(11)}$-holostadien-$3\beta$-ol |
| 3 | 22$\xi$-Hydroxy-holothurinogenin | Griseogenin | $\Delta^{7,9(11)}$-Holostadien-$3\beta$,$17\alpha$,$22\xi$-triol |
| 4 | 24,25-Dehydro-holothurinogenin | — | $\Delta^{7,9(11),24}$-Holostatrien-$3\beta$,$17\alpha$-diol |
| 5 | Holothurinogenin | — | $\Delta^{7,9(11)}$-Holostadien-$3\beta$,$17\alpha$-diol |
| 6 | 17-Desoxy-holothurinogenin | Seychellogenin | $\Delta^{7,9(11)}$-Holostadien-$3\beta$-ol |
| 7 | 25-Hydroxy-17-desoxy-holothurinogenin | Koellikerigenin | $\Delta^{7,9(11)}$-Holostadien-$3\beta$,25-diol |
| 8 | 25-Methoxy-holothurinogenin | Praslinogenin | 25-Methoxy-$\Delta^{7,9(11)}$-holostadien-$3\beta$,$17\alpha$-diol |
| 9 | 25-Methoxy-17-desoxy-holothurinogenin | Ternaygenin | 25-Methoxy-$\Delta^{7,9(11)}$-holostadien-$3\beta$-ol |

**Neoholothurinogenines**

| | | | |
|---|---|---|---|
| 10 | 12β-Methoxy-7,8-dihydro-22,25-epoxy-holothurinogenin | — | 22,25-Epoxy-12β-methoxy-$\Delta^{9(11)}$-holosten-3β,17α-diol |
| 11 | 12β-Methoxy-7,8-dihydro-17-desoxy-22,25-epoxy-holothurinogenin | — | 22,25-Epoxy-12β-methoxy-$\Delta^{9(11)}$-holosten-3β-ol |
| 12 | 12β-Methoxy-22-hydroxy-7,8-dihydro-holothurinogenin | — | 12β-Methoxy-$\Delta^{9(11)}$-holosten-3β,17α,22-triol |
| 13 | 12β-Methoxy-7,8-dihydro-24,25-dehydro-holothurinogenin | — | 12β-Methoxy-$\Delta^{9(11),24}$-holostadien-3β,17α-diol |
| 14 | 12β-Methoxy-7,8-dihydroholothurinogenin | — | 12β-Methoxy-$\Delta^{9(11)}$-holosten-3β,17α-diol |
| 15 | 12β,25-Dimethoxy-7,8-dihydroholothurinogenin | — | 12β,25-Dimethoxy-$\Delta^{9(11)}$-holosten-3β,17α-diol |
| 16 | 22,25-Epoxy-12α-methoxy-17-desoxy-7,8-dihydroholothurinogenin | — | 22,25-Epoxy-12α-methoxy-$\Delta^{9(11)}$-holosten-3β-ol |
| 17 | 12-Hydroxy-7,8-dihydro-24,25-dehydro-holothurinogenin | — | $\Delta^{9(11),24}$-Holostadien-3β,12α,17α-triol |

**Stichopogenines**

| | | | |
|---|---|---|---|
| 18 | 7,9(11)-Tetrahydro-5,8,24-hexadehydro-holothurinogenin | Stichopogenin A2 | $\Delta^{5,8,24}$-Holostatrien-3β,17α-diol |
| 19 | 7,9(11)-Tetrahydro-5,8-tetradehydro-25-hydroxyholothurinogenin | Stichopogenin A4 | $\Delta^{5,8}$-Holostadien-3β,17α,25-triol |

Holothurin has been shown to have haemolytic and cytolytic properties. It is considered to be one of the most potent saponin haemolysins known. In some concentrations it is lethal to animals and plants, inhibits the growth of certain protozoa, modifies the normal development of sea urchin eggs, possesses antimicrobial and antitumoural properties, retards pupation in the fruit fly, and inhibits regeneration processes in planariae. Some of its properties are shown in Table XV.

The effects of various preparations of holothurin on the peripheral nervous system have been the object of a number of studies by Friess and his group at the Naval Medical Research Institute in Bethesda, Maryland. These workers demonstrated that the holothurin prepared by Chanley *et al.* (1955), in concentrations of $9.8 \times 10^{-3}$ M, caused a decrease in the height of the propagated potential without reduction of the conduction velocity in the desheathed sciatic nerve of the frog. This change was concentration-dependent and independent of pH, at least between pH 7.6–8.1 at $1.95 \times 10^{-3}$ M. It was completely irreversible. A similar change was produced in the single fibre–single node of Ranvier preparation (Friess *et al.*, 1959, 1960; Thron *et al.*, 1963). In concentrations of $2.5 \times 10^{-5}$–$1.0 \times 10^{-3}$ M, the toxin produced a diminution of the action current with a concomitant rise in the stimulation threshold. This change too was irreversible and was unaffected by physostigmine in concentrations 100 or 400 times greater than the holothurin concentrations.

No change was observed on histological examination of the single fibre–single node preparation following incubation with $8.7 \times 10^{-6}$ M holothurin, the lowest concentration that produced the decrease in the action current. However, Thron *et al.* (1964) found that in approximately 80% of the preparations studied the loss of nodal excitation caused by this same concentration was accompanied by a loss in basophilic, macromolecular material from the axoplasm in and near the node of Ranvier.

Using the rat phrenic nerve-diaphragm preparation, Friess *et al.* (1959, 1960) and Thron *et al.* (1964) found that holothurin A, at a concentration of $1 \times 10^{-5}$ M and above, produced an irreversible depression in the height of both the directly and indirectly elicited contractions, the latter decreasing at a slightly greater rate than the former. These concentrations also produced an acute contracture of the muscle of short duration, which was independent of the electrical stimulation. A contracture was also observed following topical application of holothurin A on frog muscle (Thron *et al.*, 1963). If the nerve-muscle preparation was pre-treated with physostigmine at concentrations in the $10^{-10}$–$10^{-9}$ M range and then challenged with holothurin A, much of the toxin's ability to block conduction at the junction was thwarted. This "protective" effect of physostigmine was lost when concentrations were increased to $7 \times 10^{-9}$ M and higher.

TABLE XV. EFFECTS OF HOLOTHURIN ON ANIMALS AND PLANTS

| Organism | Holothurin (ppm) | Effects |
|---|---|---|
| Protozoa | | |
| *Euglena gracilis* | 66 | Growth inhibition |
| *Ochiomonas malhamensis* | 22 | Growth inhibition |
| *Tetrahymena pyriformis* | 22 | Growth inhibition |
| *Amoeba proteus* | 100 | Lethal |
| *Paramecium caudatum* | 10 | Lethal |
| Coelenterata | | |
| Hydra (brown) | 10 | Lethal |
| Platyhelminthes | | |
| *Dugesia tigrina* | 0·1–100 | Inhibition of regenerative processes; lethal |
| Nemathelminthes | | |
| Nematode (Free living) | 100 | Lethal |
| Mollusca | | |
| *Planorbis* sp. | 1–10 | Lethal |
| Annelida | | |
| *Tubifex tubifex* | 10 | Lethal |
| Arthropoda | | |
| Insect, *Drosophila melanogaster* | 1000 | Retardation of pupation |
| Echinodermata | | |
| *Arbacia punctulata*, eggs | 0.001–1000 | Developmental changes |
| Chordata | | |
| Osteichthyes, *Cyprinodon baconi* | 1–100 | Lethal |
| *Carapus bermudensis* | <1–1 | Lethal |
| Amphibia, *Rana pipiens* | 10 (mg) | Lethal (mg/animal); haemolysis and increased haemopoiesis |
| Mammalia, mouse | 9 (mg) | $LD_{50}$(IV, mg/kg) |
| | 10 (mg) | MLD (IP, mg/kg) |
| Plants | | |
| Water cress, root hairs | 1000 | Suppression of development |
| Onion, root tips | 1000 | Necrosis; lethal |

From Alender and Russell, 1966. References in original table.

Friess *et al.* (1967) demonstrated that in the mammalian phrenic nerve-diaphragm preparation the crystalline holothurin A produced a potency and degree of irreversibility of the agonistic actions that was dependent on the presence of the charged sulphate residue of the toxin. The blocking actions of a selective desulphated derivative were, for the most part, reversible on washing, in contrast to the irreversible changes produced by the parent toxin.

TABLE XVI. POTENCY INDEXES FOR SAPONIN INTERACTIONS WITH THE PHRENIC NERVE-DIAPHRAGM PREPARATION

| Saponin | Concentration $M \times 10^{-4}$ | Peak[a] muscle contracture at time (min) | N-Twitch blockade[b] | | M-Twitch blockade[b] | |
|---|---|---|---|---|---|---|
| | | | Relative amplitude pre-wash, at time[c] (min) | Amplitude post-wash, at time[c] (min) | Relative amplitude pre-wash, at time[c] (min) | Amplitude post-wash, at time[c] (min) |
| Asterosaponin A | 3·37 | 139 (2·5) | 0 (6·8) | 0 (20·4) | 34 (9·8) | 21 (20·4) |
| | 1·50 | 133 (2·0) | 25 (18·4) | 10 (31·4) | 43 (18·4) | 20 (31·4) |
| + 1·0 × 10⁻⁵ M curare[d] | 2·05 | 223 (3·1) | 0 (0) | 0 (23·9) | 17 (8·5) | 0 (23·9) |
| Asterosaponin B | 3·00 | 161 (3·2) | 42 (19·7) | 23 (29·9) | 16 (19·7) | 28 (29·9) |
| | 1·53 | 196 (2·8) | 36 (17·3) | 41 (26·1) | 27 (17·3) | 31 (26·1) |
| Holothurin B | 0·25 | 28 (363)[e] | 42 (22·0) | 4 (36·3) | 48 (22·0) | 19 (36·3) |

[a] Expressed relative to M-twitch control amplitude = 100, for isotonic work mode.

[b] Amplitudes expressed relative to respective control twitch heights = 100.

[c] Total elapsed time (min) after addition of saponin to the tissue bath.

[d] Curarized to extinction of the N-twitch at a bath level of $1·0 \times 10^{-5}$ M d-tubocurarine chloride.

[e] No peak contracture observed at this saponin concentration, but rather a steady increase in contracture with time both pre-wash and post-wash.

From Friess et al., 1967.

TABLE XVII. STRUCTURAL PROPERTIES OF PURIFIED ECHINODERM SAPONINS

| Saponin | Animal source | Presence of $-OSO_3^-Na^+$ group | Sugars per molecule | | Empirical formula | Molecular weight (average) |
|---------|---------------|----------------------------------|---------------------|---|-------------------|----------------------------|
| Holothurin A | *Actinopyga agassizi* Selenka | Yes | D-Xylose | 1 | $C_{50-52}H_{81-85}O_{25-26}SNa$ | 1159 |
| | *H. vagabunda* | | D-Glucose | 1 | | |
| | *H. lubrica* | Yes | 3-O-Methyl-D-glucose | 1 | $C_{55}H_{99}O_{29}SNa$ | 1279 |
| | | Yes | D-Quinovose | 1 | | |
| Holothurin B | *H. vagabunda* | Yes | D-Xylose | 1 | $C_{45}H_{75}O_{20}SNa$ | 991 |
| | *H. lubrica* | Yes | D-Quinovose | 1 | | |
| Asterosaponin A | *Asterias amurensis* Lütken | Yes | D-Quinovose | 2 | $C_{50-52}H_{88-90}O_{25-26}SNa$ | 1141 |
| | | | D-Fucose | 2 | | |
| Asterosaponin B | *Asterias amurensis* Lütken | Yes | D-Quinovose | 2 | $C_{56}H_{101}O_{32}SNa$ | 1341 |
| | | | D-Fucose | 1 | | |
| | | | D-Xylose | 1 | | |
| | | | D-Galactose | 1 | | |

From Friess *et al.*, 1967. References in original table.

Lethality determinations in mice indicated that the signs preceding death were the same for the two substances but that the parent toxin provokes these signs at a very much accelerated rate. Subsequently, Friess and his colleagues (1970) studied the action of holothurin A elaborated by *Actinopyga agassizi* Selenka and its neutral desulphated derivative on the cat superior cervical ganglia. Both saponins produced irreversible inactivation of the ganglion with the holothurin A being the more active.

Friess *et al.* (1968), using the rat phrenic nerve-diaphragm preparation, compared holothurin A from the sea cucumber *A. agassizi*, holothurin B from *H. vagabunda* and *H. lubrica*, and asterosaponins A and B from the sea star *Asterias amurensis* Lutken. They also summarized the structural properties of these saponins (Tables XVI and XVII).

In summarizing various of their studies Friess *et al.* (1970) note that the most obvious functional similarity from data with the cervical ganglia of the cat, the peripheral neuromuscular junction of the rat and the medullated nerve nodes of the frog was the possession of cholinergic sub-systems at some anatomical level, chiefly within the excitable membranes of conducting and junctional structures. They felt a common target for holothurin A action was "the cholinergic receptor population triggered by acetylcholine ion ($ACh^+$) in the associated hydrolase enzyme AChE, or in the enzyme choline acetylase responsible for resynthesis of $ACh^+$".

The action of holothurin A on the electrical properties of the squid axon were studied by DeGroof and Narahashi (1974). They found that external application of the toxin to the axon produced an irreversible depolarization of the membrane to nearly zero potential, while internal perfusion caused a biphasic depolarization of the membrane. The time course of the depolarization was much shorter with the internal application than with the external application. They postulated that their results demonstrated an increase in resting sodium permeability as one of the mechanisms underlying the depolarization by holothurin A.

## C.    *Clinical Problem*

In 1965, the late Dr. Carl Hubbs of the Scripps Institute of Oceanography called in consultation following an accident involving a student working with *Acanthaster planci*, who had inadvertently slipped and fallen, landing forceably with his left hand impaled upon the sea star. Twenty minutes after the injury the patient had intense pain over the palm of the left hand, "shooting pains" up the volar aspect of the forearm, weakness, nausea, vertigo and tingling in the fingertips. There were at least ten puncture wounds over the hand and some of them were bleeding freely. Dr. Hubbs was not sure if some of the spines had not broken off in the wound. I suggested that the patient put his hand in cold vinegar/water and be admitted to the hospital emergency room. On arrival there 15 min later, the pain was less intense and the nausea had somewhat subsided. Unfortunately, the patient was given 100 mg meperidine hydrochloride and 5 min later he was vomiting. (In my opinion the vomiting was due to the medication and most of the other symptoms to hyperventilation.) The patient was placed on cold vinegar/water and occasional aluminum acetate soaks over the next two days and all symptoms and signs, including the mild oedema, slowly resolved. Several broken spines were removed from the puncture wounds. Four days following the accident the patient complained of burning and itching over the left palm. Examination revealed a scaly, erythematous dermatitis. Topical corticosteroids were used but two days later the patient had to be placed on systemic corticosteroid therapy for the dermatitis. It cleared in six days.

A second episode involving *A. planci* was related to me by W. L. Orris in 1974. The patient had immediate, severe, burning pain and localized oedema. These responded to aluminium acetate soaks and corticosteroids. According to Endean (1964), the puncture wounds produced by *Asthenosoma periculosum* give rise to immediate and sometimes acute pain but few other symptoms or signs. The discharge from *Marthasterias glacialis* is said to cause oedema of the lips (Giunio, 1948). Halstead (1980) notes that contact with venomous sea stars causes "a painful wound accompanied by redness, swelling, numbness, and possible paralysis". It is known that allergic dermatitis can occur following extensive contact with these animals.

Although few people appear to enjoy eating sea stars, it is known that cats fed on *Solaster papposus* die (Parker, 1881) and that marine animals exposed to the release of the saponin or saponin-like toxin in the water will react in avoidance or, as in the case of a confined area, will die.

Stings by the pedicellariae of certain sea urchins are well documented (Russell, 1965a; Halstead, 1965; Cleland and Southcott, 1965). Fujiwara (1935) experienced severe pain, syncopy, respiratory distress, partial paralysis of the lips, tongue and eyelids, and weakness of the muscles of phonation and of the extremities following a stinging by seven or eight pedicellariae from *Toxopneustes pileolus*. Mortensen (1943) experienced severe pain of several hours duration at the site of a stinging by *Tripneustes gratilla*. The sting of a single globiferous pedicellaria from *T. gratilla* was found to be equal in pain severity to that experienced following a bee sting. Swelling appeared around the puncture wounds within minutes of the stinging and a red wheal of a centimetre in diameter soon developed. Subsequent stingings during the following two-year period resulted in a more severe reaction. In one instance the wheal was 12 cm in diameter and persisted for 8 h. In none of these experiences were there any systemic manifestations (Alender and Russell, 1966; Russell, 1971b). It might be concluded that pedicellariae stings give rise to immediate pain, localized swelling and redness, and an aching sensation in the involved part. Other findings might include those described by Fujiwara (1935) and reviewed by Halstead (1965).

As previously noted, the secondary spines of *E. calamaris*, *E. diadema*, *A. varium*, *A. thetidis* and the primary oral spines of *P. bursarium* are said to have a venom gland and are capable of envenomation (Alender and Russell, 1966). However, case reports on verified stingings are almost non-existent and in the several known to the author it is not possible to decide whether the pain, "dizziness" and minimal localized swelling were due to a venom or to the effects of a simple puncture wound complicated by hyperventilation.

The primary spines of almost 50 species of sea urchins have been implicated in injuries to man. Urchins of the family Diadematidae are particularly troublesome because of their long length and fragility. When these break off in a puncture wound they can be difficult to find and remove. I have attended injuries in which I have had to remove more than a dozen broken spine tips. With some species there is no giveaway dark colour around the puncture wound and finding the broken spines is not easy. Although the fragments of some spines will dissolve in tissue and cause no difficulties, others can give rise to granulomatous reactions, some of which may need to be removed surgically. Still others may migrate through the foot or hand without causing complications. While working in our laboratory, Dr. C. B. Alender had the tip of a spine from *T. gratilla* pass through his left hand. It had dissolved slightly but was still recognizable. It required two years to migrate through his hand. Occasionally, spines will lodge against a nerve or bone and cause complications requiring surgical intervention. Secondary infections from spine injuries are relatively rare.

It has long been known that the ovaries of sea urchins were toxic, and perhaps lethal (Loisel 1903, 1904). Halstead (1980) notes that the gonads of *Paracentrotus lividus*, *Tripneustes ventricosus* and *Centrechinus antillarium* are poisonous. Poisonings by the ingestion of certain sea cucumbers, however, are not uncommon and have occurred frequently in the South Pacific, Philippines, Japan, China and Southeast Asia. According to Yamanouchi (personal communication, 1958) the most often implicated holothurian species are *Holothuria atra*, *Holothuria axiologa*, *Stichopus variegatus* and *Thelenota ananas*. The symptoms and signs are usually of short duration and without serious sequelae. Pruritus with mild swelling and redness of the hands has been reported following the handling of some sea cucumbers. Acute conjunctivitis has been observed in persons who have swum in waters polluted with the tissue discharge of sea cucumber Cuvierian organs (Russell, 1965a).

# XI. Mollusca

Molluscs are unsegmented invertebrates having a mantle which often secretes a calcareous shell, a ventral muscular foot used for locomotion, a reduced coelom, an open circulatory system, and a radula or tongue-like organ (absent only in the bivalves). Jaws are present in some species and respiration is usually by means of gills. There are approximately 80 000 species of molluscs, of which about 85 have been implicated in poisoning to man or are known to be toxic under certain conditions. The majority of the venomous or poisonous species are found in three of the five classes of molluscs: Gastropoda, Pelecypoda, and Cephalopoda.

In the class Gastropoda, the univalve snails and slugs, the most dangerous members are of the genus *Conus*, of which there are perhaps 400 species. The cone shells are confined almost exclusively to tropical and subtropical seas and oceans and are usually found in shallow waters along reefs, although some of the more dangerous species are found on sandy bottoms. They range in length up to approximately 25 cm. Other toxic gastropods are found in species of *Aplysia*, *Creseis*, *Haliotis*, *Livona*, *Murex*, *Thais* and *Neptunea*. Halstead (1980) estimates that there are 33 000 living species of gastropods.

Among the poisonous Pelecypoda, which include the bivalves, clams, mussels and oysters are the Japanese callista, *Callista brevisiphonata*, and the giant clam, *Tridacna maxima*, common to some parts of the South Pacific, the oyster *Crassostrea gigas*, and the shellfishes *Dosinia japonica* and *Tapes semidecussata*. In the Cephalopods, the cuttlefishes, squids, nautilus and octopuses are venomous and there are possibly several poisonous squid. Of course, PSP can be found in members throughout the phylum Mollusca.

## A.  *Venom Apparatus*

The venom apparatus of *Conus* is thought to be homologous with the unpaired gland of Leiblein of certain of the higher gastropods (Fig. 14). It serves as an offensive weapon for the gaining of food and, to a much lesser extent, as a defensive weapon against predators. It consists of a muscular bulb, a long coiled venom duct, the radula (the radula sheath), and the radular teeth. The muscular pharynx and extensible proboscis are considered to be accessory organs.

The venom apparatus lies on the dorsal side of the cone in a cavity posterior to the rostrum. The venom is thought to be secreted in the venom duct and forced under pressure exerted by the duct and the venom bulb into the radula and thus into the lumen of the radular teeth. According to Kohn *et al.* (1960), the length of the venom duct may be 15 times the straight line distance, and ducts may be 5–10 cm in length, depending on the species.

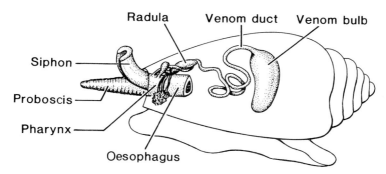

Fɪɢ. 14. Schematic representation of relationships of venom apparatus in *Conus*.

The radular teeth are passed from the radula into the pharynx and then into the proboscis. They are then thrust by the proboscis into the prey during the stinging act. The radular teeth are needle-like, from 1 to 10 mm in length, and almost transparent. They vary in size and shape depending on the species involved. The reader is referred to the works of Hinegardner (1958), Russell (1965a), Halstead (1965), Endean and Duchemin (1967), Freeman *et al.* (1974) for a more complete review of the structure of the venom apparatus of *Conus*.

Although the venom bulb has sometimes been implicated as the source of venom production, the works of Hermitte (1946), Hinegardner (1957) and Endean and Duchemin (1967) certainly indicate that the venom is produced along the length of the duct, with the posterior duct area contributing the most toxic portion. Venom may be found in the lumen of the bulb (Kohn *et al.*, 1960) but Endean and Duchemin (1967) suggest that this might be reflux from the posterior region of the venom duct and that the "sac-like cytoplasmic outgrowths containing spherules occasionally observed in the bulb epithelium are probably ruptured or separated from the epithelium when the musculature of the bulb contracts". The thick musculature of the venom bulb and its ability to forcibly squeeze the bulb probably account for the movement of venom along the duct and into the pharynx.

The radula is a Y-shaped organ lying anterior to the oesophagus-stomach and opening into the pharynx just anterior to the entrance of the venom duct. It produces the radula teeth. The organ is divided into three sections. The largest section, which overlies the oesophagus-stomach, may contain as many as 30 teeth in various states of development. The short arm of the gland attaches to the pharynx and contains most of the mature teeth. The third section is known as the ligament sac.

A ligament is attached to the base of each tooth and serves as a means of fixation while the tooth is in the radula sheath. The teeth are moved from the radula into the pharynx and thence into the proboscis. They are then thrust by the proboscis into the prey during the stinging act. It is not known whether this is done by a sling shot-like mechanism, or by hydrostatic pressure, or by some other means. In some species the tooth is held forcibly by the proboscis during the stinging act, while in others it is freed into the victim. When and how the venom gets into the radula teeth is not known. Some investigators have suggested that this occurs as the teeth are being transported through the pharynx or proboscis, while others feel that filling of the tooth and envenomation of the prey do not occur until the tooth is fired.

The venom apparatus of the octopus is an integrated part of the animal's digestive system. The secretions serve in prey capture and digestive function, in some ways similar to the venom glands of snakes. The apparatus consists of paired posterior salivary glands, two short (salivary) ducts which join them with the common salivary duct, paired anterior salivary glands and their ducts, the buccal mass and the mandibles, or beak (Fig. 15). The two paired salivary glands may differ markedly in their size, structure and function. The common salivary duct opens into the sub-radular organ anterior to the tongue. The paired ducts from the anterior salivary glands open into the posterior

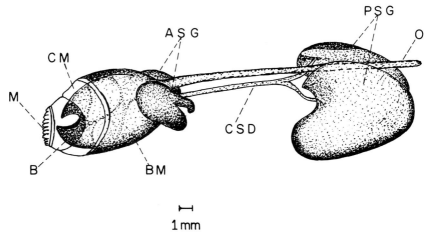

1 mm

FIG. 15. Diagrammatic sketch of the venom apparatus of an octopus. ASG, anterior salivary glands; B, beak; BM, buccal mass; CM, circular muscle; CSD, common salivary duct; O, oesophagus; M, mouth; PSG, posterior salivary glands.

pharynx. The buccal mass, or pharynx, is a muscular complex with two powerful horny jaws like an inverted parrot's beak.

Isgrove (1909) found that the venom glands of *Eledone* contained glandular secretory tubules embedded in a stroma of connective tissue. The secretory cells were columnar with a basally oriented nucleus. The secretion was formed in globules in the anterior part of these cells and passed into the lumen of the tubules. He described the secretion as a kind of mucus, containing "no ferment whatever". F. R. Modglin (see Halstead, 1965) examined the venom glands of *Octopus rubescens* and found that in the anterior salivary glands the glandular elements were lined by a simple columnar epithelium containing either peripherally or basally placed dark granules in the cytoplasm and having vesicular nuclei. He did not see ductlike structures. In some areas the lumina were dilated and contained a weblike acellular coagulum. The posterior glands consisted of fairly close-packed acini of glandular tissue that were mucous in nature, having large glandular cells with clear swollen cytoplasm and an abundance of secretory granules. He found numerous ductlike structures lined by tall, simple, columnar epithelium. These ducts were surrounded by a hyaline band similar to a basement membrane. The varying shapes of the acini and tubules suggested a convoluted arrangement and Modglin concluded that the salivary glands resembled the mucous glands of vertebrates and have merocrine type of secretion.

Gennaro *et al.* (1965) studied the salivary glands of *O. vulgaris* and found that the posterior glands consisted of branched, convoluted tubules contained within a mesenchymatous mass. Acrinar spaces were lined with secretory cells varying from cuboidal to columnar and containing secretory droplets. Within the acini were masses of pale protein material. Matus (1971), examining the posterior salivary gland of *E. cirrosa* and *O. vulgaris*, described two distinct types of epithelia lining the tubules. One type, which lined the wide lumen tubules, had a secretory function; the other did not. Gibbs and Greenaway (1978) studied the posterior salivary glands of *Hapalochaena maculosa* and *O. australis*, the latter being a non-venomous species, and found that in *H. maculosa* the gland was composed of compound tubuloacinar exocrine structures and possessed mucus secretory cells. In *O. australis* the gland was a mixture of tubular and tubuloacinar elements. They suggested that the transport of the saliva in *H. maculosa* is accomplished by muscular contraction of the gland by ciliary action and that the saliva of *O. australis* has a purely digestive function. They also suggest that there are three types of octopuses and that they represent successive evolutionary states, as far as the development of the posterior salivary glands are concerned. Their types would range from those where saliva has a purely digestive function to those which develop highly specialized toxins.

## B.  *Chemistry and Pharmacology*

The initial studies on the venoms of *Conus* indicated that the venom was white, grey, yellow or black, depending on the species involved, was viscous and had a pH range of 7·6–8·2. The active principle was non-dialysable, and its toxicity was reduced by heating or incubation with trypsin. Extracts of the venom ducts yielded homarine, gamma butyrobetaine, *N*-methyl-pyridinium, several amines, possible indole amines, 5-hydroxytryptamine, lipoproteins and carbohydrates. The lethal fraction was thought to be a protein or bound to a protein (Kohn *et al.*, 1960; Whysner and Saunders, 1966). Kohn also indicated that the various species of Conidae could be divided into those that were vermivorous, molluscivorous or piscivorous.

Endean and Rudkin (1963, 1965) and Endean *et al.* (1967) demonstrated that of 37 species studied the paralytic effects of the venoms were indeed directly related to the prey hunted. They concluded that only the piscivorous Conidae were capable of serious injuries to man. Subsequently, the Endean group found that the venoms of *Conus magus* and *C. striatus* contained conspicuous ellipsoidal granules possessing a peripheral film of polysaccharide, a sheath in which protein and lipid could be identified, and a core containing three indolyl derivatives. The average size of these granules appeared to increase as they moved from the anterior to the posterior end of the venom duct. In venom taken from the posterior end of the duct the granules had a mean long axis of 9 μm and a mean short axis of 4 μm.

Kohn *et al.* (1960) suggested that in mammals the primary deficit might be in neuromuscular transmission but the Endean group, using the venom from *C. geographus* (Whyte and Endean, 1962), *C. magus* (Endean and Izatt, 1965), and *C. striatus* (Endean *et al.*, 1967), were not able to demonstrate a specific neuromuscular deficit and attributed the effect to action on the muscle directly.

Freeman *et al.* (1974) found that the amount of venom in the ducts of *C. striatus* (Linné) is sufficient to immobilize the small fish on which it preys but not of sufficient quantity to cause serious injury to a human, They found a toxin having a molecular weight of over 10 000 with a low lethal index as compared with other *Conus* species. The toxin caused ataxia, depressed respiration leading to apnea and cardiac arrest in mammals, precipitated a block in the compound action potential of the isolated toad sciatic nerve, blocked both the directly and indirectly elicited contractions of a mammalian nerve-muscle preparation, and markedly depressed the amplitude of intracellular recorded action potentials in the rat diaphragm.

The chemical nature of the toxic fraction of *Conus* venom is not known. Various authors have reported that the active fraction is a protein, or a poison with a molecular weight of about 10 000, or a low molecular weight

TABLE XVIII. Some Venomous Species of *Conus*

| Species | Toxic to Man | Mice | Paralysis in Fishes | Molluscs | Polychaetes |
|---|---|---|---|---|---|
| **PISCIVOROUS** | | | | | |
| C. catus Hwass | X | X | X | O | O |
| C. geographus L. | X | X | X | O | O |
| C. obscurus Sowerby | X | | | | |
| C. magus L. | | X | X | + | O |
| C. stercusmuscarius L. | | X | X | + | O |
| C. striatus L. | X | X | X | + | O |
| C. tulipa L. | X | X | X | O | O |
| **MOLLUSCIVOROUS** | | | | | |
| C. ammiralis L. | | O | O | X | ? |
| C. aulicus L. | X? | X? | O | X | O |
| C. episcopus Hwass | | O | O | X | O |
| *C. marmoreus L. | X? | O | O | X | O |
| C. omaria Hwass | X? | O | O | X | O |
| C. textilis L. | X? | X | X | X | X |
| *C. tigrinus Sowerby | | O | O | X | O |
| **VERMIVOROUS** | | | | | |
| C. arenatus Hwass | | O | N | O | X |
| C. eburneus Bruguière | | N | N+ | X | X |
| C. emaciatus Reeve | | | O | O | X |
| C. flavidus Lamarck | | O | O | O | X |
| C. figulinus L. | | O | O | + | X |
| C. imperialis L. | X | O | O | O | X |
| C. leopardus (Röding) | X | | O | O | X |
| C. lividus Hwass | X | N | O | O | X |
| C. miles L. | | | O | O | X |
| C. millepunctatus Lamarck | | | O | O | X |
| C. omaria Hwass | | | O | O | X |
| C. planorbis Born | | O | + | ? | X |
| C. pulicarius Hwass | X | N | O | O | X |
| C. quercinus Solander | X | N | O | O | X |
| C. rattus Hwass | | O | X | O | X |
| C. sponsalis Hwass | X | O | O | + | X |
| C. tessulatus Born | | N | X | + | ? |
| C. virgo L. | | O | X | + | X |

\* Synonymous.
X Experimental or clinical evidence indicates toxicity.
+ May be lethal but does not produce paralysis.
? Possibly toxic, or evidence questioned or conflicting.
N Produces localized necrosis.
O Non-toxic, insofar as is known.

From Russell, 1965a.

substance, or a peptide. Proteases have been found in the anterior part of the venom duct of a number of species but not in the posterior part (Marsh, 1971). Table XVIII shows some of the venomous species of *Conus* and the kinds of animals in which they exert their deleterious effects.

Some abalone are toxic to eat. Hashimoto (1979) notes, in particular, *Haliotis sieboldi, H. japonica* and *H. discus hannai*. The toxin is concentrated in the digestive gland or liver and can be distinguished by its blue-green pigment, which is evident during the spring of the year in Japan. It is thought that the pigment, pyropheophorbide *a*, originates from chlorophyll in the seaweed on which the abalone feeds (Tsutsumi and Hashimoto, 1964).

Hashimoto *et al.* (1960) had observed that ingestion of the viscera of *Haliotis* caused dermatitis in cats and humans. On the basis of this observation they carried out experiments demonstrating the importance of photosensitization in the development of the dermatitis, suggested an assay method, and concluded that the green pigment was a chlorophyll derivative. The photosensitization was found to be dependent upon the amount of abalone consumed, time after abalone ingestion and light intensity. Hashimoto (1979) notes that the use of fluorescent pigments should be prohibited from foods and that care should be taken to see that drugs are not transformed into fluorescent substances in the body.

The hypobranchial gland of *Murex* produces a secretion which at first is colourless or yellow but on exposure to sunlight becomes brilliant violet (Tyrian purple) and gives off a strong fetid odour. The gland also produces a toxic secretion. Dubois (1903) was perhaps the first to study this toxic secretion in the puple gland of *Murex brandaris*. Roaf and Nierenstein (1907) found that the toxin gave certain colour reactions similar to those displayed by adrenalin but Jullien (1940) demonstrated that the toxin was not adrenalin, but an ester of choline. Erspamer and Dordoni (1947) showed that the active substance was not acetylcholine. They named the compound murexine.

Subsequent studies indicated that two pharmacologically active substances were present. One of these was enteramine or 5-hydroxytryptamine (Erspamer and Asero, 1952) and the other was murexine. Elementary analysis of murexine by Erspamer and Benati (1953) gave the empirical formula $C_{11}H_{18}O_3N_3$. Further study showed that murexine had the structure of β-[imidazolyl-(4)]-acrylcholine. It was thus known as urocanylcholine:

$$\begin{array}{c} H \\ | \\ N-C-H \\ | \quad \| \\ H-C \quad \| \\ \| \quad \| \qquad\qquad\qquad OH \\ N-C-CH{=}CH-COOCH_2-CH_2-N-(CH_3)_3 \end{array}$$

Blankenship *et al.* (1975) isolated murexine from the midgut of the sea hare *Aplysia californica*. Whittaker's group identified senecioylcholine in the hypobranchial gland of *Thais floridana* and acrylylcholine in *Buccinum undatum* (Whittaker and Wijesundera, 1952; Whittaker, 1960). The amount of these cholinesters in the hypobranchial gland was approximately 1–5 mg/g tissue. They are said to exhibit muscarine-like and nicotine-like activity and they cause cardiovascular changes with hypotension, increased respirations, gastric motility and secretions, and some contraction of the frog rectus muscle and guinea-pig ileum. The intravenous $LD_{50}$ of murexine in mice was 8·1–8·7 mg/kg.

Roseghini (1971) isolated dihydromurexine from the hypobranchial gland of *T. haemastoma*. It was found in high concentrations, 15–17 mg/g tissue, in the yellowish median zone of the gland. Its action on the frog rectus muscle was ten times greater than murexine or acetylcholine. Hemingway (1978) demonstrated that the muricacean gastropod *Acanthina spirata* produced a paralytic substance with high acetylcholine content in both its hypobranchial and salivary-accessory salivary gland complex. The toxin was thought to be a carboxylic ester of choline. The substances extracted from both glands were the same and were not identical to any of the choline esters reported by Keyl *et al.* (1957).

A vasodilator and hypotensive agent has been described by Huang and Mir (1972) in a salivary gland extract of the gastropod *T. haemastoma* (Clench). The extract produced behavioural changes in mice, followed by lethargy. When lethal doses were given, 43 mg fresh weight of gland/kg body weight, respirations first increased and then decreased and became shallow and death ensued. The toxin produced bradycardia and a fall in blood pressure, which was partly blocked by atropine. In the isolated rabbit heart the extract produced a decrease in rate and contraction, with a fall in heart output. It also produced contractions of the isolated guinea-pig ileum and rabbit duodenum.

Bender *et al.* (1974) found that in addition to acetylcholine, other choline esters were present in the hypobranchial glands of *A. spirata* and that *Nucella* (= *Thais*) *emarginata* contained the *N*-methyl derivative of murexine, β-[4-(1 or 3-methyl)-imidazolyl] acrylcholine, which differed pharmacologically from murexine.

The salivary poison of the gastropod *Neptunea arthritica*, which is sometimes eaten in Japan, is thought to be tetramine ($C_4H_{12}N$). It has been suggested that histamine, choline and choline ester, also found in the salivary glands of this mollusc, act synergistically with the tetramine in producing the poisoning (Asano and Itoh, 1960). Fänge (1960) notes that in *N. antiqua* the tetramine is probably responsible for almost all of the biological activity of the salivary gland extract. He found that approximately 1 % of the gland

consists of tetramine. Whittaker (1960) isolated an acetylcholine-like substance from the gastropod *Buccinum undatum*. Infra-red comparison with a
synthetic senecioylcholine indicated that the active principle was an α,β-
unsaturated derivative of isovalerylcholine.

In the viscera of the ivory shell, *Babylonia japonica*, Hashimoto *et al.*
(1967a) found a water soluble, slightly methanol and ethanol soluble, heat
labile, dialysable, ninhydrin positive and Dragendorff and biuret negative
toxin that had potent mydriatic activity. They used the mydriatic effect as an
assay method in mice. Subsequently, a partially purified fraction, which was
further purified on Sephadex G-25, was found to provoke mydriasis following
subcutaneous injection into mice at the 0·02 mg/kg body weight level. The
toxin was thought to be a polyfunctional, highly oxygenated, complex
bromo-compound having a relatively low molecular weight (Shiboto and
Hashimoto, 1970).

In 1972, Kosuge *et al.* isolated a toxic factor in crystalline form from the
mollusc *B. japonica* and determined its molecular formula, $C_{25}H_{26}N_5O_{13}Br \cdot$
$7H_2O$, with a molecular weight of 810·53. Its structure was shown as:

The toxin was named *surugatoxin* (SGTX) after Suruga Bay, where the
molluscs were taken. A crude extract (*IS-toxin*) had previously been shown to
produce ganglionic blocking of the superior cervical ganglion in the cat
(Hirayama *et al.*, 1970). Further studies showed that the toxin had a ganglion-
blocking action, that it affects neither the mammalian directly or indirectly
elicited contractions in the phrenic nerve-diaphragm preparation, that it
produced a prolonged hypotension in cats, unaffected by atropine or spinal
cord transection, and that the mode of anti-nicotinic action appeared different
from that of hexamethonium and tetraethylammonium and that it resembled
that of mecamylamine.

The Pelecypoda, scallops, oysters and clams are the principal transvectors
of paralytic shellfish poisoning, which is discussed on pages 70–94. The
genera most often involved with PSP are *Mya*, *Mytilus*, *Modiolus*, *Protothaca*,
*Spisula* and *Saxidomus*, according to Halstead (1978).

The eating of the ovaries of the Japanese callista, *Callista brevisiphonata*, has resulted in numerous cases of illness. Asano (1954) found that the ovaries of the species contained large amounts of choline but no histamine. The choline is believed to be derived from cholinesters by the action of cholinesterase, rather than from lecithin by the action of lecithinase (Hashimoto, 1979). Cats fed the shellfish showed few signs, other than hypoactivity and some loss of co-ordination. Three of nine human volunteers who ate the ovaries developed urticaria and very mild symptoms (Asano, 1954).

Venerupin poisoning was first reported in Nagai, Japan in 1889, following the ingestion of the oyster *Crassostrea gigas*. Of the 81 persons poisoned, 51 died (Halstead, 1965). A second outbreak occurred in 1941, when of 6 patients, 5 died, and from 1942 to 1950 there were 455 additional cases involving the eating of oysters and the short-necked clam *Tapes japonica* (Hashimoto, 1979). Akiba (1949) found that the toxin caused haemorrhage in the heart, lungs and viscera, with diffuse haemorrhage, necrosis and fatty degeneration of the liver. The toxin was named *venerupin*.

In 1943, Akiba initiated a series of studies on the causative agent. Neutral methanol extracts of macerated tissues of *Tapes* and *Crassostrea* were subjected to absolute ethanol. The residue was then dissolved in methanol and again precipitated with ethanol. The toxic precipitate was found to be heat stable through the pH range 3·0–8·0, and resistant to boiling for 3 h. No loss in toxicity was observed following storage in an ice box at pH 5·6 over a 20-month period. The toxin was found to be soluble in water, methanol, acetone and acetic acid. It was insoluble in benzene, ether and absolute alcohol. Certain heavy metals, picric acid and flavianic acid caused destruction of the toxin. It differed from mussel poison in that it could not be precipitated with reinecke acid. It was lethal to mice at the 0·25 mg/kg body weight level. Akiba suggested it may contain a double bond since it readily absorbed bromine. It was suggested that the toxin possessed a pyridine ring in the molecule (Akiba, 1949; Akiba and Hattori, 1949; Ishidate and Hagiwara, 1954).

The first pharmacological studies on the sea hare were carried out by Flury (1915), using *Aplysia delipans* Linnaeus and *Aplysia limacina*. He divided the secretions of *Aplysia* into three distinct types: a colourless secretion released from the surface of the animal; a terpene-like smelling, whitish, viscous material secreted from the opaline gland, and a purple substance from the purple gland. When applied to a frog's heart the secretion from the body surface caused an arrhythmia, then slowing and finally cardiac standstill. The opaline secretion caused somewhat similar changes and when it was injected into several different marine animals and frogs, all developed muscular paralysis, and several died. Testing the opaline secretion, Flury found it bitter; it produced a slight burning sensation and irritated the oral mucosa. It had no effect on the human skin but placed in the eye of a rabbit and dog it did cause immediate irritation. Flury found the secretion neutral in pH. When the purple substance was injected into frogs the animals suffered minor reflex changes and a transient paresis.

Winkler (1961) found that acetone extracts of the digestive gland of *A. californica* (Cooper) had an intraperitoneal $LD_{50}$ of approximately 30 mg sea hare tissue/kg mouse body weight. Signs included increased respiration, blanching and drooping of the ears, increased salivation, muscle fasciculations, agonal signs, ataxia, prostration and death. The extract was also lethal when given orally at approximately 12 times the intraperitoneal dose. Using column chromatography, Winkler *et al.* (1962) obtained a partially purified toxin, which they called aplysin and found that in the dog the toxin had an immediate hypotensive effect, with recovery when small doses were given. There was some initial arrhythmia followed by a slower but regular rate. In the isolated heart of the frog, aplysin caused cardiac standstill. The anterior cervical sympathetic ganglion of the cat was stimulated initially and then reversibly blocked. The frog rectus abdominis muscle responded by contracture, and the rat diaphragm–phrenic nerve neuromuscular junction was blocked; the block was antagonized by neostigmine. Rabbit intestinal muscle became spastic, with temporary cessation of peristalsis, and the spasm could be blocked by atropine. Washing of any of the preparations usually resulted in reversal of the changes.

Yamamura and Hirata (1963) isolated two bromine-containing sesquiterpenes from whole Japanese sea hares, *A. kurodai*. They named these aplysin and aplysinol:

|            | $R_1$ | $R_2$ |
|------------|-------|-------|
| Aplysin    | Br    | $CH_3$ |
| Aplysinol  | Br    | $CH_2OH$ |
| Debromoaplysin | H | $CH_3$ |

They also obtained a debromo-derivative, debromoaplysin, and subsequently a third bromo compound, diterpene aplysin-20 (Matsuda and Tomiie, 1967) was isolated. In 1952, Ando had observed a sea hare feeding on the alga *Laurencia nipponica* and conducted several experiments which indicated that steam distillates of the alga were toxic to worms and the carp. Based on this observation, Irie *et al.* (1969) extracted aplysin, debromoaplysin and aplysinol from the red alga *Laurencia okamurai*, while Waraszkiewicz and Erickson (1974) obtained aplysin from *L. nidifica*. From these observations it is suggested that the bromo toxins originate in algae.

Faulkner *et al.* (1973) and Faulkner and Stallard (1973), working with the digestive gland of *A. californica*, isolated two halogenated terpenes, including aplysin, debromoaplysin, laurenterol, johnstonol, pacifidiene and pacifidiene dichloride. Since these compounds are found in *L. pacifica*, upon which this sea hare feeds, it has been suggested that these terpenes from the mid-gut may vary, depending on the algae on which the mollusc grazes (Hashimoto, 1979).

Watson (1973) found two lethal extracts, one ether-soluble and the other water-soluble in the digestive glands of the Hawaiian sea hares *Dolabella auriculasia* (Lightfoot), *A. pulmonica* (Gould), *Styocheilus longicauda* (Quoy and Gaimard) and *Dolabrifera dolabriefa* (Rang). Homogenized glands were extracted with acetone, followed by partitioning between diethyl ether and distilled water. The ether-soluble fraction was chromatographed on a silicic acid column, while the water-soluble material was further extracted with 1-butanol. Both extracts appeared to be unaffected by brief exposure to temperatures up to 90°C, as well as to recurrent freezing and thawing over periods exceeding two years. Both were most effective at pH 2–7, and inactive above pH 8·0. Neither raw glands nor their extracts were toxic orally to mice, but both ether- and water-soluble fractions displayed effects following intraperitoneal injection. Signs following injection of the ether-soluble toxin included irritability, viciousness and severe flaccid paralysis. The water-soluble toxin, in contrast, caused "convulsions" and respiratory distress.

In a further study, Watson and Rayner (1973) observed that sublethal doses of the ether-soluble residue produced hypertension when injected intravenously into anaesthetized rats, whereas the crude water-soluble residue produced a transient hypotension, bradycardia and apnea. Dose–effect curves for more purified extracts showed actions similar to those seen with crude extracts. The hypertension produced by the ether-soluble toxin was resistant to both alpha- and beta-adrenergic blocking agents. The hypotensive effect of the water-soluble extract could not be abolished by vagotomy or pretreatment with either atropine (25 mg/kg) or Benadryl (22 mg/kg). It was concluded that both extracts may have direct effects on the contractility of vascular smooth muscle which are not mediated by alpha-adrenergic or cholinergic mechanisms.

Watson's ether-soluble toxin from *S. longicauda* was further purified by Scheuer (1975) and Kato and Scheuer (1974, 1975), who found an oily mixture of aplysiatoxin and debromoaplysiatoxin following fractionation by silica column chromatography, gel filtration, and thin layer chromatography.

| | |
|---|---|
| Aplysiatoxin | R = Br |
| Debromoaplysiatoxin | R = H |

Blankenship *et al.* (1975) confirmed the observation of Winkler *et al.* (1962) concerning the presence of choline esters in the aqueous fraction of the digestive gland of *A. californica.* They identified both acetylcholine and urocanylcholine (murexine). The latter accounted for the cholinesterase-resistant cholinomimetic activity of extracts of the gland. Kinnel *et al.* (1977) found that the "antifeedant" in *A. brasiliana* was the aromatic bromoallene, panacene. It was suggested, on the basis of studies by other workers, that the panacene is biosynthesized from a $C_{15}$ algal precursor. The provisional equation is:

Since the early contributions of Bert (1867) and Lo Bianco (1888) an impressive number of substances have been isolated from or identified in the salivary glands of various cephalopods. Many of these substances were shown to have biological activities, although these activities were not always apparent in the physiopharmacological effect of the whole toxin, and some substances either did not have a significant biological activity or the then state of knowledge did not indicate what activity was present. Finally, the amounts of the various components from the salivary glands of cephalopods were subject to such variations that it is most difficult to determine whether or not a particularly toxic substance is present in a sufficient amount to be deleterious to an envenomated animal. The importance of the synergistic effects of several of the toxic components, and of the autopharmacological response further complicated consideration of the chemistry and toxicology of this venom. It now appears that many of the substances studied were actually normal constituents of the salivary glands and not necessarily venom constituents.

A few of the substances identified from extracts of *O. vulgaris* salivary glands prior to 1964 are shown in Table XIX. The reader is referred to Russell (1965a) for the references to the identification of these substances. In the salivary glands of cephalopods, in general, tyramine, octopine, agmatine, adrenaline, noradrenaline, 5-hydroxytryptamine, L-*p*-hydroxyphenylethanolamine, histamine, dopamine, tryptophan, and certain of the 11-hydroxysteroids, polyphenols, phenolamines, indoleamines and guanidine bases have been identified.

The L-*p*-hydroxyphenylethanolamine was first described by Erspamer in 1940. It was found in extracts of the posterior salivary glands of *Octopus*

TABLE XIX. ACTIVITIES OF SALIVARY GLANDS OF *Octopus vulgaris*

| Activity | Salivary gland | |
| --- | --- | --- |
| | Posterior | Anterior |
| Tyramine oxidase | X | |
| Tryptamine oxidase | X | |
| 5-Hydroxytryptamine oxidase | X | |
| Proteolytic | X | |
| Hyaluronidase | X | X |
| Mucinolytic | X | X |
| Dopa decarboxylase | X | |
| Histamine oxidase | X | weak |
| Succinic dehydrogenase | X | X |
| Phosphatase | X | weak |
| Adenosine-triphosphatase | X | weak |
| Butyrylthiocholinesterase | X | 0 |
| Acetylthiocholinesterase | X | 0 |
| Acetylnaphtholesterase | X | X |
| Alpha naphtholase | | weak |

X Experimental evidence indicates activity.
0 Activity absent.

From Russell, 1965a. References in original table.

*vulgaris* and associated with an adrenaline-like activity. It was thought to be the precursor of hydroxyoctopamine, or L-nor-adrenaline (Erspamer, 1952). Hartman *et al.* (1960) showed that the content of the posterior salivary glands of *O. apollyon* or *O. bimaculatus* decarboxylated L-3,4-dihydroxyphenal-alanine (DOPA), DL-5-hydroxytryptophan, DL-*erythro*-3,4-dihyroxyphenyl-serine, DL-*erythro*-p-hydroxyphenylserine, DL-*m*-tyrosine, DL-*erythro*-m-hydroxyphenylserine, histidine, L-histidine, DL-*erythro*-phenylserine, 3,4-dihydroxyphenylserine, tyrosine and *m*-tyrosine. In general, the salivary glands of cephalopods were shown to contain little or no proteolytic enzymes, amylases or lipases; hyaluronidase was present in some glands.

Ghiretti (1959) purified a protein, cephalotoxin, from the posterior salivary glands of *Sepia officinalis* which he suggested was the biologically active component of the toxin. It gave positive biuret and ninhydrin reactions, and had maximum ultra-violet absorption at 276–278 nm. Four bands migrating towards the cathode were seen on starch gel electrophoresis at pH 8·5. Further purification was obtained by absorption on calcium phosphate gel at neutral pH, and three bands were obtained on electrophoresis. Treatment with trypsin resulted in the complete loss of activity. The toxin contained no cholinesterase or aminoxidase activity. Analysis of cephalotoxin from the posterior salivary gland of *Octopus vulgaris* showed: protein 74·05% (N-determination), 64·25% (biuret reaction); carbohydrates, 4·71% and hexos-amines, 5·80% (Ghiretti, 1960).

In 1949, Erspamer observed that the posterior salivary glands of *Eledone moschata* and *E. aldrovandi* contained a substance which when injected into mammals caused marked vasodilation, and produced hypotension and stimulation of certain extravascular smooth muscles. The substance was first called moschatin but was later renamed eledoisin. It was an endecapeptide having the following amino acid sequence:

Pyr–Pro–Ser–Lys–Asp(OH),–Ala–Phe–Ileu–Gly–Leu–Met–NH₂

Subsequent studies showed that eledoisin was 50 times more potent than acetycholine, histamine or bradykinin in its ability to provoke hypotension in the dog. It produced an increase in the permeability of the peripheral vessels, stimulated the smooth muscles of the gastrointestinal tract and caused an increase, which was atropine-resistant, in salivary secretions. It was easily distinguishable from the kinins and substance P (Erspamer and Anastasi, 1962). In spite of its marked pharmacological activities, the role and significance of this substance in the salivary glands of *Eledone* is not clear. It is not found in the salivary glands of *O. vulgaris* or *O. macropus*, indicating that it is not a necessary component of cephalopod toxin. It appeared that eledoisin played some part in protein synthesis in the salivary gland.

Reports of envenomation by Australian octopuses, some of which were fatal (e.g. Mabbet, 1954; Flecker and Cotton, 1955; McMichael, 1963; Hopkins, 1964), stimulated renewed interest in the venom of cephalopods. In 1964, Simon *et al.* prepared a saline extract of homogenized glands of *H. maculosa* and obtained a dialysable, heat stable product that resisted mild acid hydrolysis. This product was then studied on a number of pharmacological preparations and it was concluded that animals died in respiratory failure due to a phrenic nerve block and/or to deleterious changes at the neuromuscular junction. They also found that the extract produced bradycardia and hypotension, without remarkable changes in the electrocardiogram. A chromatographic study indicated the possible presence of octopamine and tyramine. Trethewie (1965) found that saline extracts of homogenized posterior glands of *H. maculosa* caused contraction of the guinea-pig ileum, a decrease in the isolated cat heart rate, and depression of respirations, and hypotension and cardiac failure in the intact cat. In the rat phrenic nerve–diaphragm preparation the extracts caused almost immediate cessation of the indirectly stimulated contractions and severe impairment of the directly stimulated contractions. Although quantitative studies were not done, and the dose employed appears to be rather massive, it did appear that the toxin had a marked effect on neuromuscular transmission.

Sutherland and Lane (1969) found that within four minutes of placing a live *H. maculosa* on the back of a rabbit a small bleeding puncture wound surrounded by a blanched area could be seen, the rabbit became restless, there was some exophthalmos and "one slight convulsion" with cessation of all muscular activity, other than cardiac. Cyanosis developed and death followed at 19 min. Thin layer chromatography of crude extracts of venom glands on silica gel gave nine spots, with the lethal spots having $R_f$ values of 0·20 and 0·59. Dialysis removed all toxicity, leaving two high molecular weight components. Further studies with curtain electrophoresis and acrylamide gel electrophoresis again separated the two toxins but concentration studies were difficult. The authors concluded that there were two toxins, both having molecular weights below 1000 (and perhaps below 500), that a young specimen has sufficient venom in its posterior salivary glands to cause paralysis in 750 kg of rabbits, that the gland extracts have a high concentration of hyaluronidase and that neostigmine does not reverse or reduce the toxic effects of the venom.

Freeman and Turner (1970) continued the earlier observations by studying an aqueous extract of gland tissue partially purified by filtration through Sephadex G-25. They obtained a lethal fraction that they termed maculotoxin, which they state had a similar size and molecular configuration to saxitoxin. On the basis of lethality determinations they found a close similarity between their toxin and tetrodotoxin and concluded that maculotoxin appeared to resemble tetrodotoxin more closely than it did saxitoxin. Respiratory failure on injection of maculotoxin was characterized by a loss of diaphragmatic activity which was evident before "failure of phrenic nerve volleys . . . . The vascularity of the diaphragm may permit of muscular axonal block before this is evident in the phrenic nerve".

Sutherland et al. (1970) prepared saline and water extracts of the homogenized whole glands of H. maculosa that produced paralysis and death at the 1 mg gland/2 kg rabbit level. Neither atropine and/nor neostigmine mixed with the toxin before administration altered the development of the paralysis. They demonstrated the production of antibodies to a non-toxic high molecular weight component but not for a toxic low molecular weight component, using similar techniques for both. They suggested a molecular weight below 540, which they felt accounted for the "total lack of antigenicity". As this writer has found with other venoms, the method of immunization for producing antibodies for different sized proteins varies and it is not always possible to obtain demonstrable antibodies for a small protein or a peptide using techniques best suited for large proteins. However, this might not be true for the low molecular weight protein described by these workers.

The following year it was found that maculotoxin blocked neuromuscular transmission in the isolated sciatic–sartorius nerve–muscle preparation of the toad by inhibiting the action potential in the motor nerve terminals, and that the toxin had no post-synaptic effect. It was suggested that the toxin may block action potentials by displacing sodium ions from negatively charged sites in the membrane (Dulhunty and Gage, 1971). In 1972, Croft and Howden observed one major toxin, maculotoxin, and a minor one having similar chemical properties in the venom of *H. maculosa*. Their isolation technique for the posterior salivary glands is shown in Fig. 16. The maculotoxin behaved as a cation of low molecular weight ($< 700$) and the authors felt it was chemically different from tetrodotoxin.

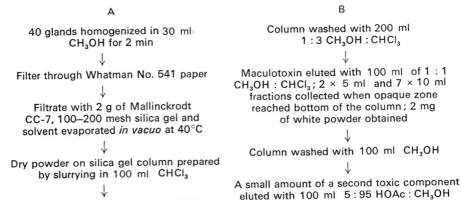

A

40 glands homogenized in 30 ml
$CH_3OH$ for 2 min

↓

Filter through Whatman No. 541 paper

↓

Filtrate with 2 g of Mallinckrodt
CC-7, 100–200 mesh silica gel and
solvent evaporated *in vacuo* at 40°C

↓

Dry powder on silica gel column prepared
by slurrying in 100 ml $CHCl_3$

↓

Column washed with 100 ml $CHCl_3$

→ B

B

Column washed with 200 ml
1 : 3 $CH_3OH$ : $CHCl_3$

↓

Maculotoxin eluted with 100 ml of 1 : 1
$CH_3OH$ : $CHCl_3$; 2 × 5 ml and 7 × 10 ml
fractions collected when opaque zone
reached bottom of the column; 2 mg
of white powder obtained

↓

Column washed with 100 ml $CH_3OH$

↓

A small amount of a second toxic component
eluted with 100 ml 5 : 95 HOAc : $CH_3OH$

FIG. 16. Isolation of maculotoxin. All fractions were tested for toxicity, for the presence of ninhydrin-positive material, and for sodium ions. The toxic fractions were evaporated *in vacuo* at 40°C (after Croft and Howden, 1972).

Although MacGinitie (1942) demonstrated the ability of the octopus to spew its salivary secretions and, indeed, R. Schweet and I obtained venom samples with Professor MacGinitie using this technique in 1951, mechanical or electrical stimulation of the animal does not seem to have found much favour among scientists. This is unfortunate because, as in the case of similar shortcomings with other animals, when extracts of a venom gland are used instead of the venom, it is quite possible that some investigators are identifying normal constituents of the venom gland rather than components of the venom. Many of the early works on marine toxins probably reflect this deficit. For this reason, the little note by Ballering *et al.* (1972) on a method for obtaining the saliva from the proboscis of *Octopus apollyon* seems important and it is hoped that researchers can be encouraged to consider this technique.

Songdahl and Shapiro (1973) fractionated buffered saline extracts of *O. dofleini* on Sephadex G-50 and G-200, studied the fractions on SDS polyacrylamide gel electrophoresis, and determined the amino acid contents. A single peak, toxic to crayfish, had a molecular weight of 23 000 ± 1000 and consisted of a single subunit with an isoelectric point of 5·2–5·3. Amino acid analysis revealed 31 aspartic, 27 glutamic, 10 lysine, and 7 arginine residues. The toxic activity was notably resistant to proteolytic destruction but was destroyed by boiling for 10 min.

Howden and Williams (1974), employing the method of Croft and Howden (1972), extracted tyramine, serotonin and histamine from the posterior salivary glands of *H. maculosa*. Their concentrations are shown in Table XX. The tyramine content is relatively low compared to that found in other species (Erspamer, 1948). The serotonin content is also comparatively low (Welsh and Moorhead, 1960), while the histamine content is consistent with that observed in other cephalopods.

TABLE XX. CONCENTRATION OF AMINES IN THE POSTERIOR SALIVARY GLANDS OF *H. maculosa*

| Sample number | Weight of glands (g) | Amount of amine present (μg/g of gland) | | |
|---|---|---|---|---|
| | | Histamine | Tyramine | Serotonin |
| 1 | 0·522 | 41 | 104 | 46 |
| 2 | 0·663 | 50 | 65 | 36 |
| 3 | 0·710 | 25 | 92 | 31 |

From Howden and Williams, 1974.

In 1975, Jarvis *et al.* again demonstrated the chemical similarities between maculotoxin and tetrodotoxin. The two toxins could be partially purified by the same techniques and were not separable by ion exclusion chromatography. The recoveries of toxicity for both poisons followed the same pattern with respect to pH. Other chemical properties were identical or similar. The authors concluded that they found no chemical evidence distinguishing the two toxins from each other. This important work has often been overlooked in reviews on the development of our knowledge on marine poisons.

Crone *et al.* (1976), using the method of Croft and Howden (1972), demonstrated that the methanol fractions contained a single toxin, maculotoxin, and again noted the similarities between maculotoxin and tetrodotoxin.

A second biologically active substance was isolated from the posterior salivary glands of *O. maculosa* by Savage and Howden (1977). It was given the name hapalotoxin and was thought to have a molecular weight of less than 700. Its estimated lethal dose appeared to be about three times that of maculotoxin, while the signs to death appear similar, although the hapalotoxin was slower in acting. Some chemical properties of the two toxins are summarized in Table XXI.

TABLE XXI. COMPARISON OF THE PROPERTIES OF HAPALOTOXIN AND MACULOTOXIN

| Property | Hapalotoxin | Maculotoxin |
|---|---|---|
| Ambient stability | Unstable | Stable |
| Behaviour on Sephadex G-10 | Retarded | Retarded |
| u.v. spectrum | End absorption | $\lambda_{max}$ 270 |
| (nm) | near 220 | $E^1_1 = 6\cdot2$ |
| TLC $R_f$, system    A | 0·20 | 0·65 |
| B | 0·08 | 0·35 |
| C | 0·30 | 0·50 |
| TLC detection $I_2$ | Positive | Positive |
| KOH/MeOH | Positive | Positive |
| ninhydrin | Negative | Positive |

From Savage and Howden, 1977.

In 1978, Sheumack *et al.* confirmed the previous observations on the likeness of maculotoxin to tetrodotoxin. Direct spectral and chromatographic comparisons showed these two toxins to be indistinguishable. This is of particular interest because here we have a poison (tetrodotoxin) which is also a venom (maculotoxin). In the former the presence of the poison is thought to be a product of metabolism, while in the latter the venom is used to immobilize and perhaps kill the prey.


## C.   Clinical Problem

A number of cones have been implicated in injuries to man, including *C. geographus, C. aulicus, C. gloria-maris, C. marmoreus, C. textilis, C. tulipa, C. striatus, C. omaria, C. catus, C. obscurus, C. imperialis, C. pulicarius, C. quercinus, C. litteratus, C. lividus* and *C. sponsalis.* The first six would seem the most dangerous, since they are said to have the highest developed venom apparatus (Halstead, 1978).

The sting often gives rise to immediate, sometimes intense, localized pain at the site of the injury. Within 5 min the victim usually notes some numbness and ischemia about the wound, although in a case seen by the author the affected area was red and tender rather than ischemic. A tingling or numbing sensation may develop about the mouth, lips and tongue, and over the peripheral parts of the extremities. Other symptoms and signs during the first 30 min following the injury include: hypertonicity, tremor, muscle fasciculations, nausea and vomiting, dizziness, increased lachrymation and salivation, weakness, and pain in the chest which increases with deep inspiration. The numbness about the wound may spread to involve a good part of the extremity or injured part. In the more severe cases, respiratory distress with chest pain, difficulties in swallowing and phonation, marked dizziness, blurring of vision and an inability to focus, ataxia, and generalized pruritus have been reported. In fatal cases, "respiratory paralysis" precedes death (Russell, 1965a).

Poisoning following the ingestion of the whelk *Neptunea arthritica* appears to be a public health problem in Hokkaido, Japan. The clinical characteristics are headache, dizziness, nausea, vomiting, weakness, ataxia, photophobia, external ocular weakness, dryness of the mouth, and, on occasions, urticaria.

Ingestion of toxic abalone produces erythema, swelling and pain over the face and neck, and sometimes the extremities, and in the more severe cases a fulminating dermatitis.

While *Murex* species have been eaten in some parts of the world without ill effects (Dubois, 1903), Plumert (1902) cites a mass poisoning at the Gulf of Trieste in which 43 persons were poisoned after eating *M. brandaris*, and five persons died. There has been some question about this incident. However, Charnot (1945) states that ingestion of *M. brandaris* may produce gastroenteritis, pruritus, convulsions and death.

Latin and medieval writers from the time of Pliny considered the sea hare *Aplysia* to be very poisonous. Their beliefs have been reviewed by Johnston (1850). In spite of the reports of ancient writers that extracts of *Aplysia* were "frequently employed to dispatch their political enemies" (Halstead, 1965), there are no recent reports of deaths from the eating of sea hares. Tasting them produces a burning sensation in the mouth and slight irritation of the oral mucosa. Handling the animals is not likely to be dangerous. However, I would suspect that it would not be safe to rub one's eyes after handling these animals.

A number of poisonings occurred in Hokkaido, Japan from the ingestion of *Callista brevisiphonata* in the early 1950s. These poisonings necessitated prohibiting the sale of the shellfish in the market place. The illness has a rapid onset, often occurring while the patient is still dining. It has been characterized as an "allergic-like" reaction, which is thought to be due to the presence of excessive choline in the ovaries. The most common findings are flushing, urticaria, and wheezing and gastrointestinal upset. It is self-limiting (Russell, 1971a,b).

Hashimoto (1979) notes a total of 542 cases of venerupin poisoning in Japan with 185 deaths. Fortunately, there have been no reported cases since 1950. It was observed following the eating of the oyster *Crassostrea gigas* or the asari *Tapes japonica*. The poisoning is characterized by a long incubation period (24–48 h, and sometimes longer; Togashi, 1943), anorexia, halitosis, nausea, vomiting, gastric pain, constipation, headache, and malaise. These findings may be followed by increased nervousness, haematemesis, and bleeding from the mucous membranes of the nose, mouth and gums. In serious cases, jaundice may be present, and petechial haemorrhages and ecchymosis may appear over the chest, neck, and arms. Leucocytosis, anaemia, and a prolonged blood-clotting time are sometimes observed. The liver is usually enlarged. In fatal poisonings, extreme excitation, delirium and coma occur. Venerupin shellfish poisoning is a public health problem in the Schizuoka and Kanagawa prefectures of Japan, where these shellfish are not eaten from January through April.

The more common types of shellfish poisoning are recognized as gastro-intestinal, allergic and paralytic. Gastrointestinal shellfish poisoning is characterized by nausea, vomiting, abdominal pain, weakness and diarrhoea. The onset of symptoms generally occurs 8–12 h following ingestion of the offending mollusc. This type of intoxication is caused by bacterial pathogens, and is usually limited to gastrointestinal signs and symptoms. It rarely persists for more than 48 h.

Allergic or erythematous shellfish poisoning is characterized by an allergic response, which may vary from one individual to another. The onset of symptoms and signs occurs 30 min–6 h after ingestion of the mollusc to which the individual is sensitive. The usual presenting signs and symptoms are diffuse erythema, swelling, urticaria and pruritus involving the head and neck and then spreading to the body. Headache, flushing, epigastric distress, and nausea are occasional complaints. In the more severe cases, generalized oedema, severe pruritus, swelling of the tongue and throat, respiratory distress, and vomiting sometimes occur. Death is rare but persons with a known sensi-tivity to shellfish should avoid eating all molluscs. The sensitizing material appears more capable of provoking a serious autopharmacological response than most known sensitizing proteins.

Paralytic shellfish poisoning is known variously as gonyaulax poisoning, paresthetic shellfish poisoning, mussel poisoning, or mytilointoxication. Pathognomonic symptoms develop within the first 30 min following ingestion of the offending mollusc. Paraesthesia, described as tingling, burning or numb-ness, is noted first about the mouth, lips and tongue; it then spreads over the face, scalp and neck, and to the finger-tips and toes. Sensory perception and proprioception are affected to the point that the individual moves inco-ordinately and in a manner similar to that seen in another, more common form of intoxication. Ataxia, incoherent speech or aphonia are prominent signs in severe poisonings.

The patient complains of dizziness, tightness of the throat and chest and some pain on deep inspiration. Weakness, malaise, headache, increased salivation and perspiration, thirst, and nausea and vomiting may be present. The pulse is usually thready and rapid; the superficial reflexes are often absent and the deep reflexes may be hypoactive. If muscular weakness and respiratory distress grow progressively more severe during the first 8 h, death may ensue. If the victim survives the first 10–12 h, the prognosis is good. Death is usually attributed to "respiratory paralysis" (Russell, 1965a; 1971a,b).

Among the cephalopods that have been implicated in bites on humans are *Hapalochaena* (= *Octopus*) *maculosa, Octopus australis, O. lunulatus, O. dofleini, O. vulgaris, O. apollyon, O. bimaculatus, O. macropus, O. rubescens, O. fitchi, O. flindersi, Ommastrephes sloani pacificus, Eledone moschata, E. aldrovandi* and *Sepia officinalis.*

Reports on serious bites by cephalopods, particularly in Australian waters, initiated more definitive studies on the venoms of these molluscs and more careful descriptions of the poisoning. The bite of most octopuses results in a small puncture wound; it appears to bleed more freely than one would expect from a similar non-envenomized traumatic wound. Pain is minimal, and in the two cases seen by the author it was described as no greater than that which would have been produced by a sharp pin. The area around the wound is first blanched but then becomes erythematous and in severe envenomations may become haemorrhagic. Tingling and numbness about the wound site are not uncommon complaints. Swelling is usually minimal immediately following the injury but may develop 6–12 h later. Muscle fasciculations have been noted following *H. maculosa* bites (Sutherland and Lane, 1969). Localized pruritus sometimes occurs over the oedematous area. "Light-headedness" of several hours' duration and weakness were reported in both cases observed by us; there were no other systemic symptoms or signs, and the wounds healed without complications (Russell, 1965a).

In the case reported by Flecker and Cotton (1955), bitten by *H. maculosa*, the patient complained of dryness in the mouth and difficulty in breathing following the bite, but no localized or generalized pain. Subsequently, breathing became more laboured, swallowing became difficult, and the patient began to vomit. Severe respiratory distress and cyanosis developed, and the victim expired. The findings at autopsy were negative. Subsequently, Cleland and Southcott (1965) reviewed the literature on cephalopod bites and noted several unreported bites on humans. Additional cases have been described by Sutherland and Lane (1969) and Snow (1970).

Poisoning following the eating of cephalopods appears to be very rare. Kawabata *et al.* (1957) describe 92 patients who became critically ill following the ingestion of *Ommastrephes sloani pacificus*. The patients developed nausea, vomiting, abdominal pain, diarrhoea, fever, headache, chills and weakness. Three patients developed paralysis and three others had convulsions. In another series of squid poisoning there were 598 cases with five deaths. The species most likely involved were *Octopus vulgaris* Lamarck and *O. dofleini* (Wulker). Intoxications took place between June to mid-September.

In none of these cases was there any evidence of bacterial poisoning, eating occurred in most instances within several hours of capture. The squid were eaten raw or cooked, while the octopuses were eaten after boiling. No evidence of histamine nor other known putrefactive amines were found in test samples.

## XII. Acknowledgements

Some of the data presented in this contribution have not heretofore been reported. They are taken from studies supported under U.S. National Institutes of Health grants GM29277 and NS1744.

I wish to express my appreciation to the following colleagues, each of whom has given advice during the preparation of this manuscript: L. Provasoli, D. A. Hessinger, W. R. Kem, Y. Shimizu, G. P. Clemons, S. L. Friess, S. Konosu, D. A. Thompson, Y. Maluf, L. J. R. Taylor, W. W. Carmichael, G. Bakus and F. S. Russell. I also wish to thank G. Childs for assistance with some of the drawings and photographs and S. Dobson for her unending editorial assistance.

## XIII. References

Abita, J.-P., Chicheportiche, R., Schweitz, H. and Lazdunski, M. (1977). Effects of neurotoxins (veratridine, sea anemone toxin, tetrodotoxin) on transmitter accumulation and release by nerve terminals in vitro. *Biochemistry* **16**, 1838.

Ackermann, D. (1922). Über die Extraktstoffe von *Mytilus edulis*. I. Mitteilung. *Z. Biol.* **74**, 67.

Aguilar-Santos, G. and Doty, M. S. (1968). Chemical studies on three species of the marine algal species *Caulerpa*. *In* "Drugs from the Sea" (H. D. Freudenthal, ed.), p. 173. Marine Technical Society, Washington, D.C.

Akiba, T. (1949). Study on poisoning by the short-necked clam and its poisonous substances. *Nissin Igaku* **36**, 1.

Akiba, T. and Hattori, Y. (1949). Food poisoning caused by eating asari (*Venerupis semidecussata*) and oyster (*Ostrea gigas*) and studies on the toxic substance, venerupin. *Jpn. J. exp. Med.* **20**, 271

Alam, M., Trieff, N. M., Ray, S. M. and Hudson, J. E. (1975). Isolation and partial characterization of toxins from dinoflagellate *Gymnodinium breve* Davis. *J. pharm. Sci.* **64**, 865.

Alcala, A. C. and Halstead, B. W. (1970). Human fatality due to ingestion of the crab *Demania* sp. in the Philippines. *Clin. Tox.* **3**, 609–611.

Alender, C. B. and Russell, F. E. (1966). Pharmacology. *In* "Physiology of Echinodermata" (R. A. Boolootian, ed.), p. 529. Interscience, New York.

Alender, C. B., Feigen, G. A. and Tomita, J. T. (1965). Isolation and characterization of sea urchin toxin. *Toxicon* **3**, 9.

Alsen, C. and Reinberg, T. (1976). Characterization of the pharmacological and toxicological actions of two toxins isolated from the sea anemone (*Anemonia sulcata*). *Bull. Inst. Pasteur* **74**, 117.

Anderson, D. M. and Wall, D. (1978). Potential importance of benthic cysts of *Gonyaulax tamarensis* and *G. excavata* in initiating toxic dinoflagellate blooms. *J. Phycol.* **14**, 224–234.

Ando, Y. (1952). Sea hare feeding behavior. *Kagaku* **22**, 87.

Anisimov, M. M., Fronert, E. B., Kuznetsova, T. A. and Elyakov, G. B. (1973). The toxic effect of triterpene glycosides from *Stichopus japonicus* Selenka on early embryogenesis of the sea urchin. *Toxicon* **11**, 109.

Anisimov, M. M., Shcheglov, V. V., Stonik, V. A., Fronert, E. B. and Elyakov, G. B. (1974). The toxic effect of cucumarioside C from *Cucumaria fraudatrix* on early embryogenesis of the sea urchin. *Toxicon* **12**, 327.

Aristotle. See (1912). "The Works of Aristotle. V. De Partibus Animalium; De Incessu Animalium et De Motu Animalium; De Generatione Animalium." (Transl. by W. Olge, A. S. L. Farquharson and A. Platt). Oxford.

Arndt, W. (1925). Über die gifte der plattwürmer. *Verhandl. Deut. zool. Ges.* **30**, 135.

Arndt, W. (1928). Die Spongern als kryptotoxische Tiere. *Zool. Jahrb.* **45**, 343.

Arndt, W. (1943). Polycladen und maricole tricladen als gifttrager. *Mém. Estud. Mus. Zool. Univ. Coimbra* **148**, 1.

Arndt, W. and Manteufel, P. (1925). Die turbellarien als träger von giften. *A. Morphol. Oekol. Tiere* **3**, 344.

Arnold, H. L., Jr, Grauer, F. H. and Chu, G. W. T. C. (1959). Seaweed dermatitis apparently caused by marine alga—clinical observations. *Proc. Hawaii Acad. Sci.* **34**, 18–19.

Asano, M. (1954). Occurrence of choline in the shellfish, *Callista brevisiphonata* Carpenter. *Tohoku J. agric. Res.* **6**, 147.

Asano, M. and Itoh, M. (1960). Salivary poison of a marine gastropod, *Neptunea arthritica* Bernardi, and the seasonal variation of its toxicity. *Ann. N.Y. Acad. Sci.* **90**, 647.

Association of Official Analytical Chemists (1975). Paralytic shellfish poison biological method. *In* "Official Methods of Analysis". 12th edn, p. 319. Association of Official Analytical Chemists, Washington, D.C.

Attaway, D. H. and Ciereszko, L. S. (1974). *In* "Proc. 2nd Int. Coral Reef Symp. 1." (A. M. Cameron, ed.), p. 497. Great Barrier Reef Committee, Brisbane, Australia.

Audebert, C. and Lamoureaux, P. (1978). Eczéma professionnel du marin pêcheur par contact de bryozoaires en baie de Seine (premiers cas français 1975–1977). *Ann. Derm. Venereol.* (Paris) **105**, 187–192.

Autenrieth, H. F. (1833). "Uber das Gift der Fische". 287 pp. C. F. Osiander, Tubingen.

Bacq, Z. M. (1936). Les poisons de némertiens. *Bull. Cl. Sci., Acad. R. Belg.* (5) **22**, 1072.

Bacq, Z. M. (1937). L' "amphiporine" et la "némertine" poisons des ver némertiens. *Arch. intern. Physiol.* **44**, 190.

Baden, D. G., Mende, T. J. and Block, R. E. (1979). Two similar toxins isolated from *Gymnodinium breve*. *In* "Toxic Dinoflagellate Blooms" (D. L. Taylor and H. H. Seliger, eds), p. 327. Elsevier/North-Holland, New York.

Baden, D. G., Mende, T. J., Lichter, W. and Wellham, L. (1981). Crystallization and toxicology of T34: a major toxin from Florida's red tide organism (*Ptychodiscus brevis*). *Toxicon* **19**, 455–462.

Bagnis, R. (1970). A case of coconut crab poisoning. *Clin. Tox.* **3**, 585–588.

Bakus, G. J. (1969). Energetics and feeding in shallow marine waters. *Int. Rev. gen. exp. Zool.* **4**, 275–369.

Bakus, G. J. (1981). Chemical defense mechanisms on the Great Barrier Reef, Australia. *Science* **211**, 497–499.

Bakus, G. J. and Green, G. (1974). Toxicity in sponges and holothurians: a geographic pattern. *Science* **185**, 951.

Bakus, G. J. and Thun, M. (1979). Bioassays on the toxicity of Caribbean sponges. *Biol. Spongiaries* **291**, 417–422.

Ballantine, D. and Abbott, B. C. (1957). Toxic marine flagellates, their occurrence and physiological effects on animals. *J. gen. Microbiol.* **16**, 274.

Ballering, R. B., Jalving, M. A., Ventresca, D. A., Hallacher, L. E., Tomlinson, J. T. and Wobber, D. R. (1972). Octopus envenomation through a plastic bag via a salivary proboscis. *Toxicon* **10**, 245.

Banner, A. H. (1959). A dermatitis-producing alga in Hawaii: preliminary report. *Hawaii med. J.* **19**, 35–36.

Banner, A. H. (1967). Marine toxins from the Pacific, I. Advances in the investigation of fish toxins. *In* "Animal Toxins" (F. E. Russell and P. R. Saunders, eds), p. 157, Pergamon Press, Oxford.

Banner, A. H. and Stephens, B. J. (1966). A note on the toxicity of the horseshoe crab in the Gulf of Thailand. *Nat. Hist. Bull. Siam Soc.* **21**, 197–203.

Barnes, J. H. (1960). Observations on jellyfish stingings in North Queensland. *Med. J. Aust.* **(2)**, 993.

Barnes, J. H. (1967a). Extraction of cnidarian venom from living tentacle. *In* "Animal Toxins" (F. E. Russell and P. R. Saunders, eds), pp. 115–129. Pergamon Press, Oxford; see also (1967) *Toxicon*, **4**, 292 (Abst.).

Barnes, J. H. (1967b). Marine stingers. Recognition and first aid treatment. Queensland Hlth. Ed. Counc., Brisbane, Hlth. Ed. Publ. **(114)**.

Baslow, M. H. (1969). "Marine Pharmacology." 286 pp. Williams and Wilkins, Baltimore, Maryland.

Bates, H. A. and Rapoport, H. (1975). A chemical assay for saxitoxin, the paralytic shellfish poison. *J. agric. Food Chem.* **23**, 237.

Baxter, E. H. and Marr, A. G. M. (1969). Sea wasp (*Chironex fleckeri*) venom: lethal, haemolytic and dermonecrotic properties. *Toxicon* **7**, 195.

Baxter, E. H. and Marr, A. G. M. (1974). Sea wasp (*Chironex fleckeri*) antivenine: neutralizing potency against the venom of three other jellyfish species. *Toxicon* **12**, 223.

Baxter, E. H. and Marr, A. G. M. (1975). Sea wasp toxoid: an immunizing agent against the venom of the box jellyfish, *Chironex fleckeri. Toxicon* **13**, 423.

Baxter, E. H., Walden, N. B. and Marr, A. G. (1972). Fatal intoxication of rabbits, sheep and monkeys by the venom of the sea wasp (*Chironex fleckeri*). *Toxicon* **10**, 653.

Bender, J. A., DeRiemer, K., Roberts, T. E., Rushton, R., Boothe, P., Mosher, H. S. and Fuhrman, F. A. (1974). Choline esters in the marine gastropods *Nucella emarginata* and *Acanthina spirata*: a new choline ester tentatively identified as *N*-methylmurexine. *Comp. gen. Pharmac.* **5**, 191–197.

Beress, L., Beress, R. and Wunderer, G. (1975). Purification of three polypeptides with neuro- and cardiotoxic activity from the sea anemone *Anemonia sulcata. Toxicon* **13**, 359.

Bergmann, F., Parnas, I. and Reich, K. (1964). The action of the toxin of *Prymnesium parvum* Carter on the guinea-pig ileum. *Brit. J. Pharmacol.* **22**, 47–55

Bergmann, W. (1949). Comparative biochemical studies on the lipids of marine invertebrates with special reference to the sterols. *J. mar. Res.* **8**, 137–76.

Bergmann, W. (1962). Sterols: their structures and distribution. *In* "Comprehensive Biochemistry" (M. Florkin and H. S. Mason, eds) pp. 103–162, 3. Academic Press, New York and London.

Bergmann, W. and Domisky, I. I. (1960). Sterols of some invertebrates. *Ann. N.Y. Acad. Sci.* **90**, 906.

Bergmann, W. and Feeney, R. J. (1951). Contributions to the study of marine products XVI. Chondrillasterol. *J. org. Chem.* **13**, 738.

Bernheimer, A. W. and Avigad, L. S. (1976). Properties of a toxin from the sea anemone *Stoichactis helianthus*, including specific binding to sphingomyelin. *Proc. natn. Acad. Sci. USA* **73**, 467.

Bernheimer, A. W. and Avigad, L. S. (1978). A cholesterol-inhibitable cytolytic protein from the sea anemone *Metridium senile*. *Biochim. Biophys. Acta* **541**, 96.

Berquist, P. R. (1978). "Sponges." 268 pp. Univ. California Press, Berkeley, California.

Berquist, P. R. and Hartman, W. D. (1969). Free amino acid patterns and the classification of the Demospongia. *Mar. Biol.* **3**, 347–368.

Berquist, P. R. and Hogg, J. J. (1969). Free amino acid patterns in Demospongiae: a biological approach to sponge classification. *Cah. Biol. Mar.* **10**, 205–220.

Bert, P. (1867). Memoire sur la physiologie de la seiche. *Mem Soc. Sci. Phys. Nat. Bordeaux* **5**, 115–138.

Bishop, C. T., Anet, E. F. L. J. and Gorham, P. R. (1959). Isolation and identification of the fast-death factor in *Microcystis aeruginosa* NRC-1. *Can. J. Biochem. Physiol.* **37**, 453.

Bisset, N. G. (1976). Hunting poisons in the North Pacific region. *Lloydia* **39**, 87–124.

Blankenship, Langlais, P. J. and Kittredge, J. S. (1975). Identification of a cholinomimetic compound in the digestive gland of *Aplysia californica*. *Comp. Biochem. Physiol.* **51C**, 129–137.

Blanquet, R. S. (1972). A toxic protein from the nematocysts of the scyphozoan medusa *Chrysaora quenquecirrha*. *Toxicon* **10**, 103.

Blumenthal, K. M. and Kem, W. R. (1980). Structure-function relationships in *Cerebratulus* toxin B-IV. *In* "Natural Toxins" (D. Eaker and T. Wadstrom, eds), pp. 487–482. Pergamon Press, Oxford.

Bodeanu, N. and Usurelu, M. (1979). Dinoflagellate blooms in Romanian Black Sea coastal waters. *In* "Toxic Dinoflagellate Blooms" (D. L. Taylor and H. H. Seliger, eds), pp. 151–154. Elsevier/North-Holland, New York.

Bolton, B. L., Bergner, A. D., O'Neill, J. J. and Wagley, P. F. (1959). Effect of a shell-fish poison on end-plate potentials. *Bull. Johns Hopkins Hosp.* **105**, 233.

Bonnevie, P. (1948). Fisherman's "Dogger bank itch" an allergic contact-eczema due to the coralline *Alcyonidium hirsutum*. *Acta Allergol.* **1**, 40.

Boolootian, R. A. (ed.) (1966). "Physiology of Echinodermata." Interscience, New York.

Bordner, J., Thiessen, W. E., Bates, H. A. and Rapoport, H. (1975). The structure of a crystalline derivative of saxitoxin. *J. Am. chem. Soc.* **97**, 6008.

Bottard, A. (1889). "Les Poissons Venimeux." 198 pp. Doin, Paris

Brieger, L. (1888). Zur Kenntniss des Tetanin und des Mytilotoxin. *Virchows Arch.* **112**, 549

Brongersma-Sanders, M. (1957). Mass mortality in the sea. *In* "Treatise on Marine Ecology and Paleoecology" (J. W. Hedgpeth, ed.), 1, *Geol. Soc. Amer., Mem.* 67, Waverly Press, Baltimore.

Burke, J. M., Marichisotto, J., McLaughlin, J. J. A. and Provasoli, L. (1960). Analysis of the toxin produced by *Gonyaulax catenella* in axenic culture. *Ann. Rev. N.Y. Acad. Sci.* 90, 837.

Burnett, J. W. and Calton, G. J. (1977). The chemistry and toxicology of some venomous pelagic coelenterates. *Toxicon* 15, 177–196.

Burnett, J. W. and Goldner, R. (1969). The effect of sea nettle (*Chrysaora quinquecirrha*) nematocyst toxin on rat cardiovascular system. *Proc. Soc. exp. Biol.* 132, 353.

Burnett, J. W., Stone, J. H., Pierce, L. H. *et al.* (1968). A physical and chemical study of sea nettle nematocysts and their toxin. *J. invest. Derm.* 51, 330.

Burreson, B. J., Christophersen, C. and Scheuer, P. J. (1975). Cooccurrence of a terpenoid isocyanide-formamide pair in the marine sponge *Halichondria* sp. *J. Am. chem. Soc.* 97, 201.

Burreson, B. J., Moore, R. E. and Roller, P. P. (1976). Volatile halogen compounds in the alga *Asparagopsis toxiformis* (Rhodophyta). *J. agric. Food Chem.* 24, 856–861.

Burton, M. (1961). Porifera. *In* "The Encyclopedia of the Biological Sciences" (P. Gray, ed.), pp. 825–826. Reinhold, New York.

Calton, G. J., Burnett, J. W., Rubinstein, H. and Heard, J. (1973). The effect of two jellyfish toxins on calcium ion transport. *Toxicon* 11, 357.

Cardellina, J. H., II, Marner, F. J. and Moore, R. E. (1979). Seaweed dermatitis: structure of lyngbyatoxin A. *Science* 204, 193.

Cariello, L., Zanetti, L. and Rathmayer, W. (1980). Isolation, purification and some properties of suberitine, the toxic protein from the marine sponge, *Suberites domuncula. In* "Natural Toxins" (D. Eaker and T. Wadström, eds), pp. 631–636. Pergamon Press, Oxford.

Carlé, J. S. and Christophersen, C. (1980). Dogger bank itch. The allergen is (2-hydroxyethyl) dimethyl-sulfoxonium ion. *J. Am. chem. Soc.* 102, 5107.

Carlé, J. S., Thybo, H. and Christophersen, C. (1982). Dogger bank itch (3). Isolation, structure determination and synthesis of a hapten. *Contact Derm.* 8, 43–47.

Carmichael, W. W. and Gorham, P. R. (1978). Anatoxins from clones of *Anabaena flos-aquae* isolated from lakes of western Canada. *Mitt. Int. Verein Limnol.* 21, 285–289.

Carmichael, W. W., Biggs, D. F. and Peterson, M. A. (1979). Pharmacology of anatoxin-a, produced by the freshwater cyanophyte *Anabaena flos-aquae* NRC-44-1. *Toxicon* 17, 229–236.

Carrilo, N. C. (1892). *Boletin de Soc. Geogr. Lima* 2, 16.

Carter, N. (1938). New or interesting algae from brackish water. *Arch Protistenk.* 90, 1.

Chanley, J. D., Kohn, S. K., Nigrelli, R. R. and Sobotka, H. (1955). Further chemical analysis of holothurin, the saponin-like steroid from the sea-cucumber Part III. *Zoologica* 40, 99.

Chanley, J. D., Perlstein, J., Nigrelli, R. F. and Sobotka, H. (1960). Further studies on the structure of holothurin. *Ann. N.Y. Acad. Sci.* 90, 202.

Chanley, J. D., Mezzetti, T. and Sobotka, H. E. (1966). The holothurinogenins. *Tetrahedron* 22, 1857.

Charnot, A. (1945). La toxicologie au Maroc. *Mém. Soc. Sci. nat. Maroc* (47), 86.

Chevallier, A. and Duchesne, E. A. (1851). Mémoire sur les empoisonnements par les huitres, les moules, les crabes, et par certains poissons de mer et de rivière. *Ann. Hyg. pub.* **46**, 108.

Cheymol, J. (1965). De dos substancias biomarinas inhabitoras neuromusculares: tetrodotoxina y saxitoxina. *Archos Fac. Med. Madrid* **8**, 151.

Cimino, G., DeStefano, S., Fenical, W., Lin, G. H. Y. *et al.* (1975). Zonaroic acid from the brown seaweed *Dictyopteris undulata* ( = zonarioides). *Experientia* **31**, 1250.

Clark, H. L. (1921). The echinoderm fauna of Torres Strait: its composition and its origin. Carnegie Inst. Washington, Publ. **214**, 155–156.

Cleland, J. B. (1913). Injuries and diseases of man in Australia attributable to animals (except insects). *J. trop Med. Hyg.* **16**, 25–31.

Cleland, J. B. (1916). "Injuries and Diseases of Man in Australia Attributable to Animals, Except Those Due to Snakes and Insects" p. 267. 6th Rept, Govt. Bur. Microbiol. Dept. Pub. Hlth, N.S.W. (Australia).

Cleland, J. B. and Southcott, R. V. (1965). "Injuries to Man from Marine Invertebrates in the Australian Region", p. 195. Commonwealth of Australia, Canberra.

Clemons, G. P., Pham, D. V. and Pinion, J. P. (1980a). Insecticidal activity of *Gonyaulax* (Dinophyceae) cell powders and saxitoxin to the German cockroach. *J. Phycol.* **16**, 305.

Clemons, G. P., Pinion, J. P., Bass, E., Pham, D. V., Sharif, M. and Wutoh, J. G. (1980b). A hemolytic principle associated with the red-tide dinoflagellate *Gonyaulax monilata*. *Toxicon* **18**, 323–326.

Cohen, S. S. (1963). Sponges, cancer chemotherapy and cellular aging. *Perspect. Biol. Med.* **6**, 215.

Collins, M. (1978). Algal toxins. *Microbiol. Rev.* **42**, 725.

Connell, C. H. and Cross, J. B. (1950). Mass mortality of fish associated with protozoan *Gonyaulax* in the Gulf of Mexico. *Science* **112**, 359–363.

Cooper, H. S. (1880). What Bêche-de-mer is, how it is caught and what is done with it. *In* "H. S. Cooper, Coral Lands", Vol. II, pp. 106–125. London.

Cormier, S. M. and Hessinger, D. A. (1980a). Cnidocil apparatus: sensory receptor of *Physalia* nematocytes. *J. Ultrastruct. Res.* **72**, 13–19.

Cormier, S. M. and Hessinger, D. A. (1980b). Cellular basis for the tentacle adherence in the Portuguese man-of-war (*Physalia physalis*). *Tissue and Cell* **12**, 713–721.

Croft, J. A. and Howden, M. E. H. (1972). Chemistry of maculotoxin: a potent neurotoxin isolated from *Hapalochlaena maculosa*. *Toxicon* **10**, 645.

Croft, J. A. and Howden, M. E. H. (1974). Isolation and partial characterization of a steroidal saponin from the starfish *Patiriella calcar*. *Comp. Biochem. Physiol.* **48B**, 535.

Crone, H. D. and Keen, T. E. B. (1969). Chromatographic properties of the hemolysin from the cnidarian *Chironex fleckeri*. *Toxicon* **7**, 79.

Crone, H. D., Leake, B., Jarvis, M. W. and Freeman, S. E. (1976). On the nature of "maculotoxin," a toxin from the blue-ringed octopus (*Hapalochlaena maculosa*). *Toxicon* **14**, 423–426.

Culver, P. and Jacobs, R. S. (1981). Lophotoxin: a neuromuscular acting toxin from the sea whip (*Lophogorgia rigida*). *Toxicon* **19**, 825–830.

Dafni, Z. and Gilberman, E. (1972). Nature of initial damage to Erlich ascites cells caused by *Prymnesium parvum* toxin. *Biochim. biophys. Acta* **255**, 380.

Das, N. P., Lim, H. S. and Teh, Y. F. (1971). Histamine and histamine-like substances in the marine sponge, *Suberites inconstans*. *Comp. gen. Pharmac.* **2**, 473–475.

De Laubenfels, M. W. (1932). The marine and freshwater sponges of California. *Proc. U.S. natn. Mus.* **81**, 1, (Publ. 2927).

DeGroof, R. C. and Narahashi, T. (1974). The effects of holothurin A on the resting membrane potential and conductance of squid axon. *Eur. J. Pharmacol.* **36**, 337–346.

De Mendiola, B. R. (1979). Red tide along the Peruvian coast. *In* "Toxic Dinoflagellate Blooms". (D. L. Taylor and H. H. Seliger, eds), 183–194. Elsevier/North-Holland, New York.

Devlin, J. P. (1974). Isolation and partial purification of hemolytic toxin from sea anemone *Stoichactis helianthus*. *J. pharm. Sci.* **63**, 1478.

Doig, M. T., III and Martin, D. F. (1973). Physical and chemical stability of ichthyotoxins produced by *Gymnodinium breve*. *Environ. Lett.* **3**, 279.

Droop, M. (1954). A note on the isolation of a small marine alga and flagellates in pure culture. *J. mar. Biol. Ass. U.K.* **33**, 511.

Drury, J. K., Noonan, J. D., Pollock, J. G. and Reid, W. H. (1980). Jellyfish sting with serious hand complications. *Injury* **12**, 66.

Dubois, R. (1903). Sur le venin de la glande à pourpre des *Murex*. *C. R. Soc. Biol.* **55**, 81.

Dubos, M., Nguyen, T. L., Lamoureux, P. *et al.* (1977). *Alcyonidinium gelatinosum* (L.) (Bryozaire) et réactions cutanées d'hypersensibilité resultats préliminaires d'une étude expérimentale. *Bull. Soc. Path. exot.* **70**, 82.

Dulhunty, A. and Gage, P. W. (1971). Selective effects of an octopus toxin on action potential. *J. Physiol.* **218**, 433–445.

Elyakov, G. B. and Peretolchin, N. V. (1970). Cucumariosid-C—a new triterpenic glycoside from *Holothuria—cucumaria fraudatrix*. *Khim. Prir. Soedin* **6**, 637.

Endean, R. (1957). The Cuvierian tubules of *Holothuria leucospilota*. *Q. J. microsc. Sci.* **98**, 455.

Endean, R. (1964). A new species of venomous echinoid from Queensland waters. *Mem. Queensland Mus.* **14**, 95–101.

Endean, R. and Duchemin, C. (1967). The venom apparatus of *Conus magus*. *Toxicon* **4**, 275–284.

Endean, R. and Henderson, L. (1969). Further studies of toxic material from nematocysts of the cubomedusan *Chironex fleckeri* Southcott. *Toxicon* **7**, 303.

Endean, R. and Izatt, J. (1965). Pharmacological study of the venom of the gastropod *Conus magus*. *Toxicon* **3**, 81.

Endean, R. and Noble, M. (1971). Toxic material from the tentacles of the cubomedusan *Chironex fleckeri*. *Toxicon* **9**, 255.

Endean, R. and Rudkin, C. (1963). Studies of the venoms of some Conidae. *Toxicon* **1**, 49.

Endean, R. and Rudkin, C. (1965). Further studies of the venoms of Conidae. *Toxicon* **2**, 255.

Endean, R., Izatt, J. and McColm, D. (1967). The venom of the piscivorous gastropod *Conus striatus*. *In* "Animal Toxins" (F. E. Russell and P. R. Saunders, eds) p. 137, Pergamon Press, Oxford; see also (1965) *Toxicon* **3**, 81; see also (1967) *Toxicon* **4**, 293.

Endean, R., Duchemin, C., McColm, D. and Fraser, E. H. (1969). A study of the biological activity of toxic material derived from nematocysts of the cubomedusan *Chironex fleckeri*. *Toxicon* **6**, 179.

Erman, A. and Neéman, I. (1977). Inhibition of phosphofructokinase by the toxic cembranolide sarcophine isolated from the soft bodied coral *Sarcophyton glaucum*. *Toxicon* **15**, 207.

Erspamer, V. (1940). Azione adrenalionsimile degli estratti di ghiandola salivare posteriore di *Octopus vulgaris* irradiati con luce ultravioletta. *Arch. Sci. Biol.* (Bologna) **26**, 443.

Erspamer, V. (1948). Active substance in the posterior salivary glands of Octopoda. I. Enteramine-like substances. *Acta pharmac. tox.* **4**, 213.

Erspamer, V. (1949). Ricerche preliminari sulla moschatina. *Experientia* **5**, 79–81.

Erspamer, V. (1952). Identification of octopamine as 1-*p*-hydroxyphenylethanolamine. *Nature* **169**, 375.

Erspamer, V. and Anastasie, A. (1962). Structure and pharmacological actions of eledoisin, the active endecapeptide of the posterior salivary glands of *Eledone*. *Experientia* **18**, 58.

Erspamer, V. and Asero, B. (1952). Identification of enteramine, the specific hormone of the entero-chromaffin cell system, as 5-hydroxytryptamine. *Nature* **169**, 300.

Erspamer, V. and Benati, O. (1953). Identification of murexine as β-(imidazoyl-(4))-acryl-choline. *Science* **117**, 161.

Erspamer, V. and Dordoni, F. (1947). Ricerche chimiche e farmacologiche sugli estratti di ghiandola ipobranchiale di *Murex trunculus, Murex brandaris* e *Tritonalia erinacea*. III. Presenze negli estratti di un nuovo derivato della colinea o di una colina omologa: la Murexina. *Arch. intern. Pharmacodyn.* **74**, 263.

Evans, M. H. (1967). Block of sensory nerve conduction in the cat by mussel poison and tetrodotoxin. *In* "Animal Toxins" (F. E. Russell and P. R. Saunders, eds), p. 97. Pergamon Press, Oxford; see also *Toxicon* **5**, 295 (Abst.).

Evans, M. H. (1968). Topical applications of saxitoxin and tetrodotoxin to peripheral nerves and spinal roots in the cat. *Toxicon* **5**, 289.

Fagerlund, V. H. M. and Idler, D. R. (1959). Marine sterols. V. Isolation of 7,24(28)-ergostadien-3β-ol from starfish. *J. Am. chem. Soc.* **81**, 401.

Fänge, R. (1960). The salivary gland of *Neptunea antiqua* (Gastropoda). *Acta Zool.* **39**, 39.

Farber, L. and Leske, P. (1963). Studies on the toxicity of *Rhodactis howesii* (Matamaw). *In* "Venomous and Poisonous Animals and Noxious Plants of the Pacific Region" (H. L. Keegan and W. V. MacFarland, eds), p. 67. Pergamon Press, Oxford.

Faulkner, D. J. and Stallard, M. O. (1973). 7-chloro-3,7-dimethyl-1,4,6-tribromo-1-octen-3-ol, a novel monoterpene alcohol from *Aplysia californica. Tetrahedron Lett.* No. 14, 1171–1174.

Faulkner, D. J. and Stallard, M. O. (1974). Chemical constituents of the digestive gland of the sea hare, *Aplysia californica*—I. Importance of diet. II. Chemical transformations. *Comp. Biochem. Physiol.* **49B**, 25, 37.

Faulkner, D. J., Stallard, M. O., Fayos, J. and Clardy, J. (1973). (3R,4S,7S)-*trans*, trans-3,7-dimethyl-1,8,8-tribromo-3,4,7-trichloro-1,5-octadiene, a novel monoterpene from the sea hare, *Aplysia californica. J. Am. chem. Soc.* **95**, 3413.

Feder, H. M. (1959). The food of the starfish *Pisaster ochraceus* along the California coast. *Ecology* **40**, 721–724.

Feder, H. M. (1963). Gastropod defensive responses and their effectiveness in reducing predation by starfishes. *Ecology* **44**, 505–512.

Feder, H. M. and Arvidsson, J. (1967). Studies on a sea-star (*Marthasterias*) *glacialis*) extract responsible for avoidance reactions in a gastropod (*Buccinum undatum*). *Arkiv Zool.* **19**, 369–379.

Feigen, G. A., Sanz, E. and Alender, C. B. (1966). Studies on the mode of action of sea urchin toxin-I. Conditions affecting release of histamine and other agents from isolated toxins. *Toxicon* **4**, 161.

Feigen, G. A., Sanz, E., Tomita, J. T. and Alender, C. B. (1968). Studies on the mode of action of sea urchin toxin-II. Enzymatic and immunological behavior. *Toxicon* **6**, 17.

Feigen, G. A., Hadji, L. and Cushing, J. E. (1974). Modes of action and identities of protein constituents in sea urchin toxin. *In* "Bioactive Compounds from the Sea" (H. J. Humm and C. E. Lane, eds), Vol. I, p. 37. Marcel Dekker, New York.

Ferlan, I. (1977). Biochemical and biological characteristics of equinatoxin isolated from *Actinia equina*. Dissertation, Univ. Ljubljana, Yugoslavia.

Ferlan, I. and Lebez, D. (1974). Equinatoxin, a lethal protein from *Actinia equina*—I. Purification and characterization. *Toxicon* **12**, 57.

Ferlan, I. and Lebez, D. (1976). Preliminary studies on the structure of equinatoxin. *Bull. Inst. Pasteur* **74**, 121.

Fingerman, M., Forester, R. H. and Stover, J. H., Jr (1953). Action of shellfish poison on peripheral nerve and skeletal muscle. *Proc. Soc. exp. Biol. Med.* **84**, 643–646.

Fish, C. J. and Cobb, M. C. (1954). Noxious marine animals of the central and western Pacific Ocean. *U.S. Fish. Wildl. Res. Rep.* **36**, 45 pp.

Flecker, H. and Cotton, B. C. (1955). Fatal bite from octopus. *Med. J. Aust.* **(2)**, 329.

Fleming, W. J. and Howden, M. E. H. (1974). Partial purification and characterization of a lethal protein from *Tripneustes gratilla*. *Toxicon* **12**, 447.

Fleming, W. J., Salathe, R., Wyllie, S. G. and Howden, M. E. H. (1976). Isolation and partial characterization of steroid glycosides from the starfish *Acanthaster planci*. *Comp. Biochem. Physiol.* **53B**, 267.

Flury, F. (1915). Über das Aplysiengift. *Arch. exp. Pathol. Pharmakol.* **79**, 250–263.

Fontaine, A. R. (1964). The integumentary mucous secretions of the ophiuroid *Ophiocomina nigra*. *J. mar. Biol. Ass. U.K.* **44**, 145.

Fraser, J. H. and Lyell, A. (1963). Dogger bank itch. *Lancet* **(1)**, 61.

Freeman, S. E. and Turner, R. J. (1969). A pharmacological study of the toxin of a cnidarian *Chironex fleckeri* Southcott. *Br. J. Pharmacol.* **35**, 510.

Freeman, S. E. and Turner, R. J. (1970). Maculotoxin, a potent toxin secreted by *Octopus maculosus* Hoyle. *Toxicol. appl. Pharmacol.* **16**, 681–690.

Freeman, S. E. and Turner, R. J. (1971). Cardiovascular effects of toxins isolated from the cnidarian *Chironex fleckeri* Southcott. *Br. J. Pharmacol.* **41**, 154.

Freeman, S. E., Turner, R. J. and Silva, S. R. (1974). The venom and venom apparatus of the marine gastropod *Conus striatus* Linne. *Toxicon* **12**, 587.

Frey, D. G. (1951). The use of sea cucumbers in poisoning fishes. *Copeia* No. 2, 175–176.

Friess, S. L. and Durant, R. C. (1965). Blockade phenomena at the mammalian neuromuscular synapse. Competition between reversible anticholinesterases and an irreversible toxin. *Toxicol. appl. Pharmacol.* **7**, 373.

Friess, S. L., Standaert, F. G., Whitcomb, E. R. *et al.* (1959). Some pharmacologic properties of holothurin, an active neurotoxin from the sea cucumber. *J. Pharm. exp. Therap.* **126**, 323.

Friess, S. L., Standaert, F. G., Whitcomb, E. R. *et al.* (1960). Some pharmacologic properties of holothurin A, a glycosidic mixture from the sea cucumber. *Ann. N.Y. Acad. Sci.* **90**, 893.

Friess, S. L., Durant, R. C., Chanley, J. D. and Fash, F. J. (1967). Role of the sulphate charge center in irreversible interactions of holothurin A with chemoreceptors. *Biochem. Pharmac.* **16**, 617.

Friess, S. L., Durant, R. C. and Chanley, J. D. (1968). Further studies on biological actions of steroidal saponins produced by poisonous echinoderms. *Toxicon* **6**, 81.

Friess, S. L., Chanley, J. D., Hudak, W. V. and Weems, H. B. (1970). Interactions of the echinoderm toxin holothurin A and its desulfated derivative with the cat superior cervical ganglion preparation. *Toxicon* **8**, 211–219.

Fujiwara, T. (1935). [On the poisonous pedicellariae of *Toxopneustes pileolus* (Lamarck).] *Annotnes zool. Jpn.* **15**, 62.

Fusetani, N. and Hashimoto, Y. (1976). Hemolysins in a green alga *Ulva pertusa*. *In* "Animal, Plant and Microbial Toxins" (A. Ohsaka, K. Hayashi and Y. Sawai, eds), Vol. 1, pp. 325–332. Plenum Publishing Corp., New York.

Fusetani, N. and Hashimoto, K. (1981). Diethyl peroxides probably responsible for *moyuku* poisoning. *Bull. Jpn. Soc. scient. Fish.* **47**, 1059–1063.

Fusetani, N., Ozawa, C. and Hashimoto, Y. (1976). Fatty acids as ichthyotoxic constituents of a green alga *Chaetomorpha minima*. *Bull. Jpn. Soc. scient. Fish.* **42**, 941.

Fusetani, N., Hashimoto, K., Mizukami, I., Kamiya, H. and Yonabaru, S. (1980). Lethality in mice of the coconut crab *Birgus latro*. *Toxicon* **18**, 694–698.

Galenus (Galen, C.) (1597). "Opera Omnia." Giunta, Venice.

Galsoff, P. S. (1948). Red tide. Progress report on the investigations of the cause of the mortality along the west coast of Florida conducted by the United States Fish and Wildlife Service and cooperating organizations. *Spec. Sci. Rep. U.S. Fish. Wildl. Serv.* **46**, 44 pp.

Garth, J. S. and Alcala, A. C. (1977). Poisonous crabs of Indo-West Pacific coral reefs, with special reference to the genus *Demania laurie*. *In* "Proc. Third Int. Coral Reef Symp", pp. 645–652. Miami, Florida.

Gasteiger, E. L., Haake, P. C. and Gergen, J. A. (1960). An investigation of the distribution and function of homarine (N-methyl picolinic acid). *Ann. N.Y. Acad. Sci.* **90**, 622.

Gazio, A. (1491). "Corona Florida Medicinae sive de Conservatione Sanitatis." Joannes & Gregorius de Gregoris de Forlivio, Venice.

Gennaro, J. F., Jr, Lorincz, A. E. and Brewster, N. B. (1965). The anterior salivary gland of the octopus (*Octopus vulgaris*) and its mucous secretions. *Ann. N.Y. Acad. Sci.* **118**, 1021.

Gentile, J. (1971). Blue-green and green algal toxins. *In* "Microbial Toxins" (S. Kadis, A. Ciegler, and S. Ajl, eds), Vol. 7, p. 27. Academic Press, New York and London.

Gershey, R. M., Neve, R. A., Musgrave, D. L. and Reichardt, P. B. (1977). A colorimetric method for the determination of saxitoxin. *J. Fish. Res. Bd. Can.* **34**, 559.

Gesner, K. (1551–1587). "Historia Animalium." C. Froschouerum, Tiguri; see also (1558) "Historiae Animalium." Zurich and (1563) "Fischbuch das ist ein Kurtze." C. Froschouer, Zurich.

Ghiretti, F. (1959). Cephalotoxin: the crab-paralysing agent of the posterior salivary glands of cephalopods. *Nature* **183**, 1192.

Ghiretti, F. (1960). Toxicity of octopus saliva against crustacea. *Ann. N.Y. Acad. Sci.* **90**, 726.

Gibbs, F. J. and Greenaway, P. (1978). Histological structure of the posterior salivary glands in the blue-ringed octopus *Hapalochlaena maculosa*. *Toxicon* **16**, 59.

Giunio, P. (1948). Otrovne ribe. *Higijena* (Belgrade) **1**, 282

Giraldi, T., Ferlan, I. and Romeo, D. (1976). Antitumor activity of equinatoxin. *Chem. Biol. Interact.* **13**, 199.

Glaser, O. C. and Sparrow, C. M. (1909). The physiology of nematocysts. *J. exp. Zool.* **6**, 361.

Graham, A. (1955). Molluscan diets. *Proc. malac. Soc.* (Lond.) **31**, 144–158.

Grauer, F. H. (1959). Dermatitis escharotica caused by a marine alga. *Hawaii med. J.* **19**, 32–34.

Grauer, F. H. and Arnold, H. L., Jr (1961). Seaweed dermatitis. *Arch. Derm.* **84**, 720–730.

Green, G. (1977). Ecology of toxicity in marine sponges. *Mar. Biol.* **40**, 207–215.

Grevin, J. (1568). "Deux livres des venins", p. 423. Christolfe Plantin, Paris.

Grindley, J. R. and Taylor, F. J. R. (1962). Red water and mass mortality of fish near Cape Town. *Nature* (Lond) **195**, 1324.

Grunfeld, Y. and Spiegelstein, M. Y. (1974). Effects of *Gymnodinium breve* toxin on the smooth muscle preparations of guinea-pig ileum. *Br. J. Pharmacol.* **51**, 67.

Guar, P. K., Anthony, R. L., Cody, T. S. *et al.* (1981). Production of a monoclonal antibody against the sea nettle venom mouse lethal factors. *Proc. Soc. exp. Biol. Med.* **167**, 374–377.

Guldager, A. (1959). Doggerbankeeksem, forsøg pa profylaktisk behandling med corticosteroider. *Ugeskr. Laeg.* **121**, 1567.

Habermehl, G. and Volkwein, G. (1968). Über Gifte der mittelmeerischen Holothurien. I. Mitteilung. *Naturwissenschaften* **55**, 83.

Habermehl, G. and Volkwein, G. (1970). Über Gifte der mittelmeerischen Holothurien. II. Die Aglyka der toxine von *Holothuria polii. Liebigs Ann. Chem.* **731**, 53.

Habermehl, G. and Volkwein, G. (1971). Aglycones of the toxins from the cuvierian organs of *Holothuria forskali* and a new nomenclature for the aglycones from Holothurioideae. *Toxicon* **9**, 319–326.

Halstead, B. W. (1964). Fish poisonings—their diagnosis, pharmacology, and treatment. *Clin. Pharmacol. Therap.* **5**, 615–627.

Halstead, B. W. (1965–1970). "Poisonous and Venomous Marine Animals of the World." 3 vols. U.S. Govt. Print. Off., Washington, D.C.

Halstead, B. W. (1978). "Poisonous and Venomous Marine Animals of the World." Revised edn, 238 pp. Darwin Press, Inc., Princeton, New Jersey.

Halstead, B. W. (1980). "Dangerous Marine Animals", 2nd edn, p. 77. Cornell Maritime Press, Centreville, Maryland.

Hand, C. (1961). Present state of nematocyst research: types, structure and function. *In* "The Biology of Hydra and of Some Other Coelenterates" (H. M. Lenhoff and W. F. Loomis, eds), p. 187. Univ. Miami Press, Coral Gables, Florida.

Hartman, W. J., Clark, W. G., Cyr, S. C., Jordon, A. L. and Liebhold, R. A. (1960). Pharmacologically active amines and their biogenesis in the octopus. *Proc. west. pharmacol. Soc.* **3**, 106.

Hashimoto, Y. (1979). "Marine Toxins and Other Bioactive Metabolites", 369 pp. Japan Scientific Society, Tokyo.

Hashimoto, Y. and Ashida, K. (1973). Screening of toxic corals and isolation of a toxic polypeptide from *Goniopora* spp. *Publ. Seto Mar. Biol. Lab.* **20**, 703.

Hashimoto, Y. and Yasumoto, T. (1960). Confirmation of saponin as a toxic principle of starfish. *Bull. Jpn. Soc. scient. Fish.* **26**, 1132–1138.

Hashimoto, Y., Naito, K. and Tsutumi, J. (1960). Photosensitization of animals by the viscera of abalones, *Haliotis* spp. *Bull. Jpn. Soc. scient. Fish.* **26**, 1216–1218.

Hashimoto, Y., Miyazawa, K., Kamiya, H. and Shibota, M. (1967a). Toxicity of the Japanese ivory shell. *Bull. Jpn. Soc. scient. Fish.* **33**, 661–668.

Hashimoto, Y., Konosu, S., Yasumoto, T., Inoue, A. and Noguchi, T. (1967b). Occurrence of toxic crabs in Ryukyu and Amami Islands. *Toxicon* **5**, 85.

Hashimoto, Y., Fusetani, N. and Kimura, S. (1969a). Aluterin: a toxin of filefish, *Alutera scripta*, probably originating from a zoantharian, *Palythoa tuberculosa*. *Bull. Jpn. Soc. scient. Fish.* **35**, 1086.

Hashimoto, Y., Konosu, S., Yasumoto, T. and Kamiya, H. (1969b). Ciguatera in the Ryukyu and Amami Islands. *Bull. Jpn. Soc. scient. Fish.* **35**, 316.

Hemingway, G. T. (1978). Evidence for a paralytic venom in the intertidal snail, *Acanthina spirata* (Neogastropoda: Thaisidae). *Comp. Biochem. Physiol.* **60C**, 79–81.

Henri, V. and Kayalof, E. (1906). Étude des toxines contenues dans les pédicellaires chez les Oursins. *C.R. Soc. Biol.* **60**, 884.

Hermitte, L. C. (1946). Venomous marine molluscs of the genus *Conus*. *Trans. Roy. Soc. trop. Med. Hyg.* **39**, 485.

Hessinger, D. A. and Lenhoff, H. M. (1976). Membrane structure and function. Mechanism of hemolysis induced by nematocyst venom: roles of phospholipase and direct lytic factor. *Arch. Biochem. Biophys.* **173**, 603.

Hillard (1938). Science and scientists. *Med. Arts* **3**, 1.

Hinegardner, R. T. (1957). "Morphology of the venom apparatus of *Conus*." Thesis. Univ. S. California, Los Angeles.

Hinegardner, R. T. (1958). The venom apparatus of the cone shell. *Hawaii med. J.* **17**, 533.

Hines, K. and Lane, C. E. (1962). Amino acid composition of active peptides of *Physalia* toxin. *Fed. Proc.* **21**, 53.

Hirata, Y., Uemura, D., Ueda, K. and Takano, S. (1979). Several compounds from *Palythoa tuberculosa* (Coelenterata). *Pure appl. Chem.* **51**, 1875.

Hirayama, H., Gohgi, K., Urakawa, N. and Ikeda, M. (1970). A ganglionic-blocking action of the toxin isolated from Japanese ivory shell (*Babylonica japonica*). *Jpn. J. Pharmac.* **20**, 311–312.

Holthuis, L. B. (1968). Are there poisonous crabs? *Crustaceana* **15**, 215–222.

Hopkins, D. G. (1964). Venomous effects and treatment of octopus bite. *Med. J. Aust.* **(3)**, 81.

Hoppe-Seyler, F. A. (1933). Über das Homarin, eine bisher unbekannte tierische Base. *Hoppe-Seyler Z.* **222**, 105–115.

Hornell, J. (1917). A new protozoan cause of widespread mortality among marine fishes. *Bull. Madras Fish. Dept.* **11**, 53–56.

Howden, M. E. H. and Williams, P. A. (1974). Occurrence of amines in the posterior salivary glands of the octopus *Hapalochlaena maculosa* (Cephalopoda). *Toxicon* **12**, 317.

Howell, J. F. (1953). *Gonyaulax monilata* sp. nov., the causative dinoflagellate of a red tide on the east coast of Florida in August–September, 1951. *Trans. Am. microsc. Soc.* **72**, 153.

Huang, C. L. and Mir, G. N. (1972). Pharmacological investigation of salivary gland of *Thais haemastoma* (Clench). *Toxicon* **10**, 111–117.

Huber, C. S. (1972). The crystal structure and absolute configuration of 2,9 diacetyl-9-azabicyclo(4,2,1)non-2,3-ene. *Acta Crystallogr.* **B28**, 2577.

Hughes, E., Gorham, P. and Zehnder, A. (1958). Toxicity of a unialgal culture of *Microcystis aeruginosa*. *Can. J. Microbiol.* **4**, 225.

Hulme, F. E. (1895). "Natural History of Lore and Legend", 350 pp. London.

Hyman, L. H. (1940). "The Invertebrates: Protozoa through Ctenophora", Vol. I, 726 pp. McGraw-Hill, New York.

Hyman, L. H. (1955). "The Invertebrates: Echinodermata", 763 pp. McGraw-Hill, New York.

Ikawa, M. and Taylor, R. (1973). Choline and related substances in algae. *In* "Marine Pharmacognosy. Action of Marine Biotoxins at the Cellular Level" D. Martin and G. Padilla, eds), p. 203. Academic Press, New York and London.

Ikegami, S., Kamiya, Y. and Tamura, S. (1972). Isolation and characterization of spawning inhibitors in ovaries of the starfish, *Asterias amurensis*. *Agric. Biol. Chem.* **36**, 1087.

Imai, M. and Inoue, K. (1974). The mechanism of the action of prymnesium toxin on membranes. *Biochim. Biophys. Acta* **352**, 344–348.

Inoue, A., Noguchi, T., Konosu, S. and Hashimoto, Y. (1968). A new toxic crab, *Altergatis floridus*. *Toxicon* **6**, 119–123.

Irie, T., Suzuki, M. and Hayakawa, Y. (1969). Isolation of aplysin, debromoaplysin, and aplysinol from *Laurencia okamurai* Yamada. *Bull. chem. Soc. Jpn.* **42**, 843–844.

Isgrove, A. (1909). Eledone. *Mem. Liverpool mar. Biol. Comm.* **18**, 105.

Ishidate, M. and Hagiwara, H. (1954). [On the poisonous substance isolated from asari (*Venerupis semidecussata*) and oyster.] *Exp. Rept., Dept. nat. Sci.*, (Univ. Tokyo) **10**, 43.

Jakowska, S. and Nigrelli, R. F. (1960). Antimicrobial substances from sponges. *Ann. N.Y. Acad. Sci.* **90**, 913.

Jarvis, M. W., Crone, H. D., Freeman, S. E. and Turner, R. J. (1975). Chromatographic properties of maculotoxin, a toxin secreted by *Octopus (Hapalochlaena) maculosus*. *Toxicon* **13**, 177.

Jennings, R. K. and Aker, R. F. (1970). "The Protistan Kingdom", 120 pp. Van Nostrand Reinhold, New York.

Johnson, H. M., Frey, P. A., Angellotte, R., Campbell, J. E. and Lewis, K. L. (1964). Haptenic properties of paralytic shellfish poisoning. *Proc. Soc. exp. biol. Med.* **117**, 425.

Johnston G. (1850). "An Introduction to Conchology. Or, Elements of the Natural History of Molluscous Animals", 614 pp. J. Van Voorst, London.

Jullien, A. (1940). Variations dans le temps de la teneur des extraits de glande à pourpre en substances actives sur le muscle de sangsue. *C. R. Soc. Biol.* **133**, 524.

Kaiser, E. and Michl, H. (1958). "Die Biochemie der tierischen Gifte", 258 pp. Franz Deuticke, Wien.

Kao, C. Y. (1967). Comparison of the biological actions of tetrodotoxin and saxitoxin. *In* "Animal Toxins" (F. E. Russell and P. R. Saunders, eds), pp. 109–114. Pergamon Press, Oxford.

Kashman, Y., Fishelson, L. and Neéman, I. (1973). *N*-acyl-2-methylene-β-alanine methyl esters from the sponge *Fasciospongia cavernosa*. *Tetrahedron* **29**, 3655.

Kato, Y. and Scheuer, P. J. (1974). Aplysiatoxin and debromoaplysiatoxin, constituents of the marine mollusk *Stylocheilus longicauda* (Quoy and Gaimard, 1824). *J. Am. chem. Soc.* **96**, 2245–2246.

Kato, Y. and Scheuer, P. J. (1975). The aplysiatoxins. *Pure appl. Chem.* **42**, 1–14.

Kawabata, T., Halstead, B. W. and Judefind, T. F. (1957). A report of a series of recent outbreaks of unusual cephalopod and fish intoxication in Japan. *Am. J. trop. Med. Hyg.* **6**, 935–939.

Keen, T. E. B. and Crone, H. D. (1969a). The hemolytic properties of extracts of tentacles from the cnidarian *Chironex fleckeri*. *Toxicon* **7**, 55–63.

Keen, T. E. B. and Crone, H. D. (1969b). The dermonecrotic properties of extracts from the tentacles of the cnidarian *Chironex fleckeri*. *Toxicon* **7**, 173–180.

Kelecom, A., Daloze, D. and Tursch, B. (1976). Chemical studies of marine invertebrates. *Tetrahedron* **32**, 2313,

Keleti, G., Sykora, J. L., Lippy, E. C. and Shapiro, M. A. (1979). Composition and biological properties of lipopolysaccharides isolated from *Schizothrix calcicola* (Ag.) Gomont (cyanobacteria). *Appl. environ. Microbiol.* **38**, 471–474.

Kellaway, C. H. (1935a). The action of mussel poison on the nervous system. *Aust. J. exp. Biol. Med. Sci.* **13**, 79.

Kellaway, C. H. (1935b). Mussel poisoning. *Med. J. Aust.* (1), 399.

Kem, W. R. (1969). "A chemical investigation of nemertine toxins." Ph.D. Thesis, University of Illinois, Urbana, Illinois.

Kem, W. R. (1971). A study of the occurrence of anabaseine in *Paranamertes* and other nemertines. *Toxicon* **9**, 23.

Kem, W. R. (1973). Biochemistry of nemertine toxins. *In* "Marine Pharmacognosy. Action of Marine Biotoxins at the Cellular Level" (D. F. Martin and G. M. Padilla, eds), p. 37. Academic Press, New York and London.

Kem, W. R. (1976). Purification and characterization of a new family of poly-peptide neurotoxins from the heteronemertine *Cerebratulus lacteus* (Leidy). *J. Biol. Chem.* **251**, 4184.

Kem, W. R. and Blumenthal, K. M. (1978). Polypeptide cytolysins and neurotoxins isolated from the mucus secretions of the heteronemertine *Cerebratulus lacteus* (Leidy). *In* "Toxins. Animal, Plant and Microbial" (P. Rosenberg, ed), pp. 509–516. Pergamon Press, Oxford.

Kem, W. R., Abbot, B. C. and Coates, R. M. (1971). Isolation and structure of a hoplonemertine toxin. *Toxicon* **9**, 15.

Kem, W. R., Blumenthal, K. M. and Doyle, J. W. (1980). Cytotoxins of some marine invertebrates. *In* "Natural Toxins" (D. Eaker and T. Wadstrom, eds), pp. 157–163. Pergamon Press, Oxford.

Keyl, M. J., Michaelson, I. A. and Whittaker, V. P. (1957). Physiologically active choline esters in certain gastropods and other invertebrates. *J. Physiol.* **139**, 434.

Kim, Y. S. and Padilla, G. A. (1977). Hemolytically active components from *P. parvum* and *G. breve* toxins. *Life Sci.* **21**, 1287.

Kimura, A., Hayashi, H. and Kuramoto, M. (1975). Studies of urchi-toxins: separation, purification and pharmacological actions of toxinic substances. *Jpn. J. Pharmacol.* **25**, 109–120.

King, H. (1939). Amphiporine, an active base from the marine worm *Amphiporus lactifloreus*. *J. Am. chem. Soc.* **2**, 1365.

Kinnel, R., Duggan, A. J., Eisner, T. and Meinwald, J. (1977). Panacene: an aromatic broncalene from a sea hare (*Aplysia brasiliana*). *Tetrahedron Lett.* **44**, 3913.

Kitigawa, I., Sugawara, T. and Yoshioka, I. (1974). Structure of Holotoxin A, a major antifungal glycoside of *Stichopus japonicus selenka*. *Tetrahedron Lett.* **47**, 4111.

Kitigawa, I., Sugawara, T., Yoshioka, I. and Kunyama, K. (1976). [Holotoxins of *Stichopus* sp.] *Chem. Pharm. Bull.* **24**, 275.

Kittredge, J. S., Takahashi, F. T., Lindsey, J. and Lasker, R. (1974). Chemical signals in the sea: marine allelochemics and evolution. *Fish. Bull.* **72**, 1.

Klauber, L. M. (1956). "Rattlesnakes. Their Habits, Life Histories and Influence on Mankind", 2 vols. University of California Press, Berkeley, California.

Kleinhaus, A. L., Cranefield, P. F. and Burnett, J. W. (1973). The effects on canine cardiac Purkinje fibers of *Chrysaora quinquecirrha* (sea nettle) toxin. *Toxicon* **11**, 341.

Kofoid, C. A. (1911). Dinoflagellata of the San Diego region, IV. The genus *Gonyaulax*, with notes on its skeletal morphology and a discussion of its generic and specific characteristics. *Univ. Calif. Publ. Zool.* **8**, 187–286.

Kohn, A. J., Saunders, P. R. and Wiener, S. (1960). Preliminary studies on the venom of the marine snail *Conus. Ann. N.Y. Acad. Sci.* **90**, 706.

Konishi, K. (1970). Studies on organic insecticides. Part XII. Synthesis of nereis-toxin and related compounds. V. *Agric. biol. Chem.* **34**, 935–940.

Konosu, S. (1968). Toxic crabs in the Ryukyu and Amami Islands. *Kagaku to Seibutsu* **6**, 413.

Konosu, S., Inoue, A., Noguchi, T. and Hashimoto, Y. (1968). Comparison of crab toxin with saxitoxin and tetrodotoxin. *Toxicon* **6**, 113–117.

Kosuge, T., Zenda, H., Ochiai, A., Masaki, N. *et al.* (1972). Isolation and structure determination of a new marine toxin, surugatoxin, from the Japanese ivory shell, *Babylonia japonica. Tetrahedron Lett.* **26**, 2545.

Lal, D. M., Calton, G. J., Neéman, I. and Burnett, J. W. (1981). Characterization of *Chrysaora quinquecirrha* (sea nettle) nematocyst venom collagenase. *Comp. Biochem. Physiol.* **69B**, 529–533.

Lane, C. E. (1960). The toxin of *Physalia* nematocysts. *Ann. N.Y. Acad. Sci.* **90**, 742.

Lane, C. E. (1961). *Physalia* nematocysts and their toxin. *In* "Biology of Hydra" (H. M. Lenhoff and W. F. Loomis, eds), p. 169. Univ. Miami Press, Coral Gables, Florida.

Lane, C. E. (1967). Pharmacologic action of *Physalia* toxin. *Fed. Proc.* **26**, 1225.

Lane, C. E. and Dodge, E. (1958). The toxicity of *Physalia* nematocysts. *Biol. Bull.* **115**, 219–226.

Lane, C. E. and Larsen, J. B. (1965). Some effects of the toxin of *Physalia physalis* on the heart of the land crab, *Cardisoma guanhumi* (Latreille). *Toxicon* **3**, 69.

Larsen, J. B. and Lane, C. E. (1966). Some effects of *Physalia physalis* toxin on the cardiovascular system of the rat. *Toxicon* **4**, 199.

Larsen, J. B. and Lane, C. E. (1970). Direct action of *Physalia* toxin on frog nerve and muscle. *Toxicon* **8**, 21.

Lassabliere, M. P. (1906). Influence des injections intraveineuses de suberitine sur la resistance globulaire. *C.R. Seanc. Soc. Biol.* **61**, 600.

Lenhoff, H. M., Kline, E. S. and Hurley, R. (1957). A hydroxyproline-rich, intra-cellular, collagen-like protein of *Hydra* nematocysts. *Biochim. Biophys. Acta* **26**, 204–205.

Lentz, T. L. (1966). Histochemical localization of neurohumors in a sponge. *J. exp. Zool.* **162**, 171.

Lentz, T. L. and Barrnett, R. J. (1962). The effect of enzyme substrates and pharma-cological agents on nematocyst discharge. *J. exp. Zool.* **149**, 33.

Lentz, T. L. and Barrnett, R. J. (1965). Fine structure of the nervous system of *Hydra. Amer. Zool.* **5**, 341–356.

Levy, R. (1925). Sur les propriétés hémolytiques des pedicellaires de certains Oursins réguliers. *C.R. Acad. Sci.* **181**, 690.

Liebert, F. and Deerns, W. (1920). Ondeszoek naar de oorzaak van den vischsterfte in den polur Workumer-Nieuwland, nabij Workum. *Verh. Rapp. Rijksinst. Visscherijonderzoe* **1**, 81.

Lightner, D. V. (1978). Possible toxic effects of the marine blue-green alga, *Sperulina subsala*, on the blue shrimp, *Penaeus stylirostris. J. invert. Path.* **32**, 139–150.

Lindner, G. (1888). Über giftige miesmuscheln. *Centralbl. Zent. Bak. Parasiten* **3**, 352.

Lippy, E. C. and Erb, J. (1976). Gastrointestinal illness at Sewickley, Pa. *J. Am. Waterworks Assoc.* **68**, 606.

LoBianco, S. (1888). Notizie biologiche riguardanti specialmente il periodo di maturita sessuale degli animali del golfo di Napoli. *Mitt. Zool. Sta. Neapel* **8**, 385.

Loisel, G. (1903). Les poisons des glandes genitales. 1er note: recherches et experimentation chez l'oursin. *C. R. Soc. Biol.* **55**, 1329.

Loisel, G. (1904). Recherches sur les poisons genitaux de differents animaux. *C. R. Acad. Sci.* **139**, 227.

Lubbock, R. (1979). Chemical recognition and nematocyte excitation in a sea anemone. *J. exp. Biol.* **83**, 283.

Lubbock, R. (1980). Why are clownfishes not stung by sea anemone? *Proc. Roy. Soc. London* **B207**, 35.

Mabbet, H. (1954). Death of a skindiver. *Austral. Skin Div. Spear Fish. Dig.* (Dec.), p. 13, 17.

Maček, P. and Lebez, D. (1981). Kinetics of hemolysis induced by equinatoxin, a cytolytic toxin from the sea anemone *Actinia equina*. Effect of some ions and pH. *Toxicon* **19**, 233–240.

Macfarlane, R. D., Uemura, D. and Hirata, Y. (1980). $C_f$ plasma desorption mass spectrometry of palytoxin. *J. Am. chem. Soc.* **102**, 875.

MacGinitie, G. E. (1942). Notes on the natural history of some marine animals. *Am. Midland Naturalist* **19**, 213–214.

Macht, D. I., Brooks, D J. and Spencer, E. C. (1941). Physiological and toxicological effects of some fish muscle extracts. *Am. J. Physiol.* **133**, 372.

Mackie, A. M. and Turner, A. B. (1970). Partial characterization of a biologically active steroid glycoside isolated from starfish *Marthasterias glacialis*. *Biochem. J.* **117**, 543.

Mackie, A. M., Lasker, R. and Grant, P. T. (1968). Avoidance reactions of the mollusc *Buccinum undatum* to saponin-like surface-active substances in extracts of the starfish *Asterias rubens* and *Marthasterias glacialis*. *Comp. Biochem. Physiol.* **26**, 415.

Mackie, G. O. (1960). Studies on *Physalia physalis* (L.). Part 2. Behavior and histology. *Discovery Rep.* **30**, 371–407.

Maiti, B. C., Thomson, R. H. and Mahendra, M. (1978). Structure of caulerpin, a pigment from *Caulerpa* algae. *J. chem. Res.* **4**, 126–127.

Maretić, Z. (1975). "Zivitinje Otrovnice I Otrovne Zivitinje Jadranskog Mora", pp. 11–22. Jugoslave Akad. Znanosti Umjetnosti, Zagreb, Yugoslavia.

Maretić, Z. and Russell, F. E. (1983). Stings by the anemone *Anemonia sulcata*. *Am. J. Trop. Med. Hyg.* **32**, 891.

Maretić, Z., Russell, F. E. and Ladavac, J. (1980). Epidemic of stings by the jellyfish *Pelagia noctiluca* in the Adriatic. *In* "Natural Toxins" (D. Eaker and T. Wadstrom, eds) pp. 77–82. Pergamon Press, Oxford; see also *Toxicon* **17** (Suppl. 1), 115 (Abst.).

Margolin, A. S. (1964). A running response of *Acmaea* to sea stars. *Ecology* **45**, 191–193.

Mariscal, R. N. (1974). Nematocysts. *In* "Coelenterate Biology" (L. Muscatine and H. M. Lenhoff, eds), pp. 129–178. Academic Press, London and New York.

Marsh, H. (1971). The caseinase activity of some vermivorous Conidae. *Toxicon* **8**, 271.

Martin, D. F. and Chatterjee, A. B. (1970). Some chemical and physical properties of two toxins from the red-tide organism, *Gymnodinium breve*. *U.S. Fish. Wildl. Serv., Fish. Bull.* **68**, 433.

Martin, D. F. and Padilla, G. M. (eds) (1973). "Marine Pharmacognosy." 317 pp. Academic Press, New York and London.

Martin, D. F. and Padilla, G. M. (1974). Effect of *Gymnodinium breve* toxin on potassium influx of erythrocytes. *Toxicon* 12, 353.

Martin, E. J. (1960). Observations on the toxic sea anemone *Rhodactis howesii* (Coelenterata). *Pac. Sci.* 17, 32.

Martyr, P. (1533). "De Orbe Novo; The Eight Decades of Peter Martyr d'Anghera." J. Bebelium, Basileae; (1912) (transl. by F. A. McNutt), 2 vols. Putnam Sons, New York.

Matsuda, H. and Tomiie, Y. (1967). The structure of aplysin-20. *Chem. Commun.*, 898–899.

Matsuno, T. and Iba, T. (1966). [Studies on the saponin of sea cucumber.] *J. pharm. Soc. Jpn.* 86, 637.

Matsuno, T. and Yamanouchi, T. (1961). A new triterpenoid sapogenin of animal origin (sea cucumber). *Nature* 191, 75.

Matus, A. I. (1971). Fine structure of the posterior salivary gland of *Eledone cirrosa* and *Octopus vulgaris*. *Z. Zellforsch.* 122, 11.

McFarren, E. F., Schafer, M. L., Campbell, J. E. *et al.* (1956). Public health significance of paralytic shellfish poison. A review of literature and unpublished research. *Proc. natn. Shellfish Ass.* 47, 114.

McFarren, E. F., Schaffer, M. L., Campbell, J. E. *et al.* (1960). Public health significance of paralytic shellfish poison. *Adv. Food Res.* 10, 135.

McLaughlin, J. J. and Provasoli, L. (1957). Nutritional requirements and toxicity of two marine *Amphrolineum*. *J. Protozool.* 4, (Suppl. 1), 7.

McMichael, D. F. (1963). Dangerous marine molluscs. Proc. First Intern. Convention Life Saving Techniques, Part III, Sci, Sect., p. 86. Bull. Post Grad. Comm. Med., Univ. Sydney, Australia.

Mebs, D. and Gebauer, E. (1980). Isolation of proteinase inhibitor, toxic and hemolytic polypeptides from a sea anemone, *Stoichactis* sp. *Toxicon* 18, 97–106.

Medcof, J. C., Leim, A. H., Needler, A. B. *et al.* (1947). Paralytic shellfish poisoning on the Canadian Atlantic coast. *Bull. Fish. Res. Bd. Can.* 75, 1.

Mendes, E. G. (1963). Ulteriores experimentos sobre o principio colinergico dans pedicelarias. *Cienc. e Cult* 15, 275.

Meyer, K. F. (1953). Medical progress: food poisoning. *New Engl. med. J.* 249, 765, 804, 843.

Meyer, K. F., Sommer, H. and Schoenholz, P. (1928). Mussel poisoning. *J. prev. Med.* 2, 365.

Michaels, W. D. (1979). Membrane damage by a toxin from the sea anemone *Stoichactis helicanthus*—I. Formation of trans membrane channels in lipid bilayers. *Biochim. Biophys. Acta* 555, 67.

Michel, C. and Keil, B. (1975). Biologically active proteins in the venomous glands of the polychaetous annelid, *Glycera convoluta* Keferstein. *Comp. Biochem. Physiol.* 50B, 29–33.

Middlebrook, R. E., Wittle, L. W., Scura, E. D. and Lane, C. E. (1971). Isolation and purification of a toxin from *Millepora dichotoma*. *Toxicon* 9, 333–336.

Minale, L. (1978). Terpenoids from marine sponges. In "Marine Natural Products" (P. J. Scheuer, ed.), Vol. I, pp. 175–240. Academic Press, New York and London.

Mold, J. D., Bowden, J. P., Stanger, D. W. *et al.* (1957). Paralytic shellfish poison; VII. Evidence for the purity of the poison isolated from toxic clams and mussels. *J. Am. chem. Soc.* 79, 5235.

Möller, H. and Beress, L. (1975). Effect on fishes of two toxic polypeptides isolated from *Anemonia sulcata*. *Mar. Biol.* 32, 189–192.

Montgomery, D. H. (1967). Responses of two haliotid gastropods (*Mollusca*) *Haliotis asimilis* and *Haliotis rufescens* to the forcipulate asteroids (Echinodermata) *Pycnopodia helianthoides* and *Pisaster ochraceus*. *Veliger* **9**, 359–368.

Moore, R. E. (1977). Toxins from blue-green algae. *Bioscience* **27**, 797.

Moore, R. E. (1981). Constituents of blue-green algae. In "Marine Natural Products. Chemical and Biological Perspectives" (P. J. Scheuer, ed.), Vol. IV, pp. 1–52. Academic Press, New York and London.

Moore, R. E. and Bartolini, G. (1981). Structure of palytoxin. *J. Am. chem. Soc.* **103**, 2491–2494.

Moore, R. E. and Scheuer, P. J. (1971). Palytoxin: a new marine toxin from a coelenterate. *Science* **172**, 495.

Mori, Y., Anraku, M., Yagi, K. *et al.* (1968). On the venomous crustacea and fish from Amami-Oshima and Okinawa Islands—I. *Med. J. Kagoshima Univ.* **19**, 729.

Mortensen, T. (1928–1951). "A Monograph of the Echinoidea." 5 Vols. C. A. Reitzel, Copenhagen.

Mote, G. E., Halstead, B. W. and Hashimoto, Y. (1970). Occurrence of toxic crabs in the Palau Islands. *Clin. Tox.* **3**, 597–607.

Müller, H. (1935). Chemistry and toxicity of mussel poison. *J. Pharmacol. exp. Ther.* **53**, 67.

Mullin, C. A. (1923). Report on some polychaetous annelids; collected by the Barbados-Antigua expedition from the University of Iowa in 1918. *Univ. Iowa Stud. nat. Hist.* **10**, 39–45.

Munro, H. S. (1964). "The nature and mode of action of coelenterate toxins." Thesis, Harvard University.

Murtha, E. F. (1960). Pharmacological study of poisons from shellfish and puffer fish. *Ann. N. Y. Acad. Sci.* **90**, 820.

Murthy, S. S. N. and der Marderosian, A. (1973). In "Food-Drugs from the Sea Proceedings 1972" (L. R. Worthen, ed.), p. 181. Marine Technology Society, Washington, D.C.

Mynderse, J. S. and Moore, R. E. (1978). Toxins from blue-green algae: structures of oscillatoxin A and three related bromine-containing toxins. *J. org. Chem.* **43**, 2301.

Mynderse, J. S., Moore, R. E., Kashiwagi, M. and Norton, T. R. (1977). Anti-leukemic activity in the Oscillatoriaceae: isolation of debromoaplysiatoxin from *Lyngbya*. *Science* **196**, 538.

National Marine Fisheries Service (1977). "Oxygen Depletion and Associated Environmental Disturbances in the Middle Atlantic Bight in 1976." N.M.F.T., N.E.F.C., Sandy Hook Laboratory, Highlands, N. J., Tech. Ser. Rept. No. 3, 1–471.

Natori, T. (1972). ["Ainu and Archaeology."] Vol. I, pp. 202–204. Hokkaido Shuppan Kikaku Sentea, Sapporo.

Neéman, I., Fishelson, L. and Kashman, Y. (1974). Sarcophine—a new toxin from the soft coral *Sarcophyton glaucum* (Alcyonaria). *Toxicon* **12**, 593.

Neéman, I., Calton, G. J. and Burnett, J. W. (1980a). An ultrastructure study of the cytotoxic effect of the venoms from the sea nettle (*Chrysaora quinquecirrha*) and Portuguese man-of-war (*Physalia physalis*) on cultured Chinese hamster ovary K-1 cells. *Toxicon* **18**, 495–501.

Neéman, I., Calton, G. J. and Burnett, J. W. (1980b). Cytotoxicity and dermo-necrosis of sea nettle (*Chrysaora quinquecirrha*) venom. *Toxicon* **18**, 55–64.

Neéman, I., Calton, G. J. and Burnett, J. W. (1981). Purification of an endonuclease present in *Chrysaora quinquecirrha* venom. *Proc. Soc. exp. Biol. Med.* **166**, 374.

Newhouse, M. L. (1966). Dogger bank itch: survey of trawlermen. *Br. med. J.* **1**, 1142.

Nicander, U. (1499). "Theriaca et Alexipharmaca." Manutium, Venetiis.

Nicols, D. (1962). "Echinoderms." 200 pp. Hutchinson Univ. Library, London.

Nightingale, W. H. (1936). "Red Water Organisms: Their Occurrence and Influence Upon Marine Aquatic Animals, with Special Reference to Shellfish in Waters of the Pacific Coast." 24 pp. Argus Press, Seattle. Washington.

Nigrelli, R. F. (1952). The effects of holothurin on fish, and mice with sarcoma 180. *Zoologica* **37**, 89.

Nigrelli, R. F. and Jakowska, S. (1960). Effects of holothurin, a steroid saponin from the Bahamian sea cucumber (*Actinopyga agassizi*) on various biological systems. *Ann. N.Y. Acad. Sci.* **90**, 884.

Nigrelli, R. F. and Zahl, P. A. (1952). Some biological characteristics of holothurin. *Proc. Soc. exp. Biol. Med.* **81**, 379.

Nilsson, A. and Fänge, R. (1967). The digestive fluid of *Priapulus caudatus* (Lam.). *Comp. Biochem. Physiol.* **22**, 927–931.

Nitta, S. (1934). Über Nereistoxin, einen giftigen Bestandteil von *Lumbriconereis heteropoda* (Eunicidae). *J. Pharm. Soc. Jpn.* **54**, 648.

Norton, T. R., Shibata, S., Kashiwagi, M. and Bentley, J. (1976). Isolation and characterization of the cardiotonic polypeptide Anthopleurin-A from the sea anemone *Anthopleura xanthogrammica*. *J. pharm. Sci.* **65**, 1368.

Novak, V., Sket, D., Cankar, G. and Lebez, D. (1973). Partial purification of a toxin from tentacles of the sea anemone *Anemonia sulcata*. *Toxicon* **11**, 411.

O'Connell, M. G. (1971). "Fine structure of venom gland cells in globiferous pedicellariae from sea urchin." Thesis, California State College at Long Beach.

Olson, T. A. (1951). Toxin plankton. *In* "Proceedings in Service Training Course in Waterworks Problems", pp. 86–96. Univ. Mich. Sch. Pub. Hlth., Ann Arbor, Mich., Feb. 15–16, 1951.

Owellen, R. J., Owellen, R. G., Gorog, M. A. and Klein, D. (1973). Cytolytic fraction from *Asterias vulgaris*. *Toxicon* **11**, 319.

Padilla, G. M. and Martin, D. F. (1973). Interactions of *Prymnesium parvum* toxin with erythrocyte membranes. *In* "Tier und Pflanzengifte/Animal and Plant Toxins" (E. Kaiser, ed.), W. Goldmann, Munich; see also (1973) "Marine Pharmacognosy. Action of Marine Biotoxins at the Cellular Level" (D. F. Martin and G. M. Padilla, eds), p. 265, Academic Press, New York and London; see also (1972) *Toxicon* **10**, 532.

Padilla, G. M., Kim, Y. S. and Martin, D. F. (1974). Separation and analysis of toxins isolated from a red-tide sample of *Gymnodinium breve*. *In* "Proceedings of the First International Conference on Toxic Dinoflagellate Blooms" (Locicero, V. R., ed.), p. 249. Mass. Sci. Tech. Found., Wakefield, Mass.

Pantin, C. F. (1942). The excitation of nematocysts. *J. exp. Biol.* **19**, 294.

Parker, C. A. (1881). Poisonous qualities of the star-fish. *Zoologist* **5**, 214.

Parker, G. H. and Van Alstyne, M. A. (1932). The control and discharge of nematocysts, especially in *Metridium* and *Physalia*. *J. exp. Zool.* **63**, 329.

Parnas, I. and Abbott, B. C. (1965). Physiological activity of the ichthyotoxin from *Prymnesium parvum*. *Toxicon* **3**, 133.

Parnas, I. and Russell, F. E. (1967). Effects of venoms on nerve, muscle and neuromuscular junction. *In* "Animal Toxins" (F. E. Russell and P. R. Saunders, eds), pp. 401–415. Pergamon Press, Oxford.

Paster, Z. (1968). "Purification and properties of prymnesin, the toxin formed by *Prymnesium parvum* (Chrysomonadinae)." Ph.D. thesis, Hebrew University, Jerusalem.

Paster, Z. (1973). Pharmacognosy and mode of action of *Prymnesium*. *In* "Marine Pharmacognosy. Action of Marine Biotoxins at the Cellular Level" (D. F. Martin and G. M. Padilla, eds), p. 241. Academic Press, New York and London.

Paster, Z. and Abbott, B. C. (1969). Hemolysis of rabbit erythrocytes by *Gymnodinium breve* toxin. *Toxicon* **7**, 245.

Pawlowsky, E. N. (1927). "Gifttiere und Ihre Giftigkeit." 516 pp. G. Fisher, Jena.

Penčar, S., Ferlan, I., Cotic, L. and Schara, M. (1975). An EPR study of the hemolitic action of equinatoxin. *Period. Biol.* **77**, 149.

Penner, L. R. (1970). Bristleworm stinging in a natural environment. *Univ. Conn. Occ. Pap.* **1**, 275–280.

Pepler, W. J. and Loubser, E. (1960). Histochemical demonstration of the mode of action of the alkaloid in mussel poisoning. *Nature* **188**, 800.

Pérès, J. M. (1950). Recherches sur les pédicellaires glandulaires de *Sphaerechinus granularis* (Lamarck). *Arch. Zool. exp. gen.* **86**, 118.

Peter of Abanos (Petrus de Albano) (1472). "Tractatus de Venenes." Mantua.

Phillips, C. and Brady, W. H. (1953). "Sea Pests: Poisonous or Harmful Sea Life of Florida and the West Indies." 78 pp. Univ. Miami Press, Miami, Florida.

Phillips, J. H. (1956). Isolation of active nematocysts of *Metridium senile* and their chemical composition. *Nature* **178**, 1932.

Phisalix, M. (1922). "Animaux Venimeux et Venins", Vol. I, 656 pp.; Vol. II, 864 pp. Masson et Cie, Paris.

Pliny (Caius Plinius Secundus). See (1601) "The Histoire of the World. Commonly Called, The Natural Histoire of C. Plinius Secundus" (Transl. by P. Holland), London.

Plumert, A. (1902). Über giftige seetiere im allgemeinen und einen Fall von Massenvergiftung durch Seemuscheln im besonderen. *Arch. Schiffs-u. Tropenhyg.* **6**, 15.

Pope, E. C. (1947). Sea animals that sting and bite *Aust. Mus. Mag.* **9**, 164.

Pope, E. C. (1963). Some noxious marine invertebrates from Australian seas. Proc. First Intern. Convention Life Saving Techniques, Part III, Sci. Sect., p. 91. Bull. Post Grad. Comm. Med., Univ. Sydney.

Posner, P. and Kem, W. R. (1978). Cardiac effects of toxin A-III from the heteronemertine worm *Cerebratulus lacteus* (Leidy). *Toxicon* **16**, 343–349.

Prakash, A., Medcof, J. C. and Tennant, A. D. (1971). Paralytic shellfish poisoning in eastern Canada. *Bull. Fish. Res. Bd. Can.* **177**, 1–87.

Prinzmetal, M., Sommer, H. and Leake, C. D. (1932). The pharmacological action of "mussel poison". *J. Pharmacol. exp. Therap.* **46**, 63.

Proctor, N. H., Chan, S. L. and Taylor, A. J. (1975). Production of saxitoxin by cultures of *Gonyaulax catenella*. *Toxicon* **13**, 1.

Provasoli, L. (1978). Recent progress, an overview. *In* "Toxic Dinoflagellate Blooms" (D. L. Taylor and H. H. Seliger, eds) pp. 1–14. Elsevier/North-Holland New York.

Rama Murthy, J. and Capindale, J. G. (1970). A new isolation and structure for the endotoxin *Microcystis aeruginosa* NRC-1. *Can. J. Biochem.* **48**, 508–510.

Randall, J. E. and Hartman, W. D. (1968). Sponge feeding fishes of the West Indies. *Mar. Biol.* **1**, 216–225.

Rathmayer, W., Jessen, B. and Beress, L. (1975). Effect of toxins of sea anemones on neuromuscular transmission. *Naturwissenschaften* **62**, 538.

Read, B. E. (1939). Chinese Materia Medica: fish drugs. *Peking nat. Hist. Bull.*, 136 pp.

Reich, K. and Kahn, J. (1954). A bacteria-free culture of *Prymnesium parvum* (Chrysomonadina). *Bull. Res. Counc. Israel* **4**, 144.

Rice, N. E. and Powell, W. A. (1970). Observations on three species of jelly-fishes from the Chesapeake Bay with special reference to their toxins. I. *Chrysaora (Pactylometra) quinquecirrha. Biol. Bull.* **139**, 180.

Rich, A. C. (1882). Crayfish poisoning: a series of cases. *Liverpool Med. Chir. J.* **2**, 384.

Richet, C. (1906). De la variabilité de la dose toxique de subéritine. *C.R. Soc. Biol.* **61**, 686.

Richet, C. (1907). Anaphylaxie par la mytilo-congestine. *C.R. Soc. Biol.* **62**, 358.

Rio, G. J., Ruggieri, G. D., Stempien, M. F., Jr and Nigrelli, R. F. (1963). Saponin-like toxin from the giant sunburst starfish, *Pycnopodia helianthoides* from the Pacific Northwest. *Am Zool.* **3**, 544.

Roaf, H. E. and Nierenstein, M. (1907). Adrénaline et purpurine (reply to M. R. Dubois). *C.R. Soc. Biol.* **63**, 773.

Robson, E. A. (1972). The behaviour and neuromuscular system of *Gonactinia prolifera*, a swimming sea anemone. *J. exp. Biol.* **55**, 611–640.

Rondelet, G. (1554–1555). "Libri de piscibus marinis." M. Bonhomme, Lyons; (1554) "Universae aquatilium historiae . . ." M. Bonhomme, Lugduni; (1558) "L'histoire Entiere de Poissons." M. Bonhomme, Lyons.

Roseghini, M. (1971). Occurrence of dihydromurexine (imadazolepropionylcholine) in the hypobranchial gland of the *Thais naemastoma. Experientia* **27**, 1008.

Ruggieri, G. D. and Nigrelli, R. F. (1973). Effects of extracts of the sea star *Acanthaster planci* on the developing sea urchin. *Am. Soc. Zool.* **380**, 592–593.

Ruggieri, G. D., Nigrelli, R. F. and Stempien, R. F., Jr (1970). Some biochemical and physiological properties of extracts from several echinoderms. p. 75. 2nd Intern. Symp. Animal Toxins, Tel Aviv.

Runnegar, M. T. C. and Falconer, I. R. (1975). Isolation of toxin from a naturally-occurring algal bloom of *Microcystic-aeruginosa* ( =*Anacystis cyanea*). *Proc. Austral. Biochem. Soc.* **8**, 5.

Russell, F. E. (1965a). Marine toxins and venomous and poisonous marine animals. *In* "Advances in Marine Biology" (F. S. Russell, ed.), Vol. 3, pp. 255–384. Academic Press, London and New York.

Russell, F. E. (1965b). Venomous and poisonous marine animals and their toxins. First Inter-American Naval Res. Conf., San Juan, Puerto Rico, July 26, 37 pp.

Russell, F. E. (1966a). To be, or not to be . . . the $LD_{50}$. *Toxicon* **4**, 81–83.

Russell, F. E. (1966b). Physalia stings—a report of two cases. *Toxicon* **4**, 65.

Russell, F. E. (1967). Comparative pharmacology of some animal toxins. *Fed. Proc.* **26**, 1206–1224.

Russell, F. E. (1971a). "Poisonous Marine Animals. (Marine Toxins and Venomous and Poisonous Marine Animals)." 176 pp. T.F.H. Publications, Neptune City, N.J.

Russell, F. E. (1971b). Pharmacology of toxins of marine origin. *In* "International Encyclopedia of Pharmacology and Therapeutics" (H. Raskova, ed.), Sect. 71, Vol. II, pp. 3–114. Pergamon Press, Oxford.

Russell, F. E. (1975). Poisonous and venomous marine animals and their toxins. *Ann. N.Y. Acad. Sci.* **245**, 57–64.

Russell, F. E. (1980a). Pharmacology of venoms. *In* "Natural Toxins." (Proc. 6th Intern. Symp. Animal, Plant, Microb. Toxins, Uppsala, 1979) (D. Eaker and T. Wadstrom, eds), p. 13–21. Pergamon Press, Oxford.

Russell, F. E. (1980b). "Snake Venom Poisoning." 562 pp. Lippincott, Philadelphia.

Russell, F. S. (1953). "The Medusae of the British Isles." Cambridge University Press, Cambridge, England.

Salkowski, E. (1885). Zur Kenntniss des Giftes der Miesmuschel (*Mytilus edulis*). *Virchows Arch.* **102**, 578.

Sasner, J. J., Ikawa, M., Thurberg, F. and Alam, M. (1972). Physiological and chemical studies on *Gymnodinium breve* Davis toxin. *Toxicon* **10**, 163.

Savage, I. V. E. and Howden, M. E .H. (1977). Hapalotoxin, a second lethal toxin from the octopus *Hapalochlaena maculosa*. *Toxicon* **15**, 463.

Saville-Kent, W. (1893). "The Great Barrier Reef of Australia—Its Products and Potentialities." W. H. Allen & Co., London.

Schantz, E. J. (1960). Biochemical studies on paralytic shellfish poisons. *Ann. N. Y. Acad. Sci.* **90**, 843.

Schantz, E. J. (1963). Studies on the paralytic poisons found in mussels and clams along the North American Pacific coast. *In* "Venomous and Poisonous Animals and Noxious Plants of the Pacific Region" (H. L. Keegan and M. V. MacFarlane, eds), p. 75. Pergamon Press, Oxford.

Schantz, E. J. (1971). The dinoflagellate poisons. *In* "Microbial Toxins. Algal and Fungal Toxins" (S. Kadis, A. Ciegler and S. Ajl, eds), Vol. III, pp. 3–26. Academic Press, New York and London.

Schantz, E. J., Mold, J. D., Stanger, D. W. *et al.* (1957). Paralytic shellfish poison. VI. A procedure for the isolation and purification of the poison from toxic clam and mussel tissues. *J. Am. chem. Soc.* **79**, 5230.

Schantz, E. J., McFarren, E. F., Schaffer, M. L. and Lewis, K. H. (1958). Purified shellfish poison for bioassay standardization. *J. Assoc. Off. agric. Chem.* **41**, 160.

Schantz, E. J , Lynch, J. M., Vayvada, G., Matsumoto, K. and Rapoport, H. (1966). The purification and characterization of the poison produced by *Gonyaulax catanella* in axenic culture. *Biochemistry* **5**, 1191.

Schantz, E. J., Ghazarossian, V. E., Schnoes, H. K. *et al.* (1975). Paralytic poisons from marine dinoflagellates. *In* "Proceedings of the First International Conference on Toxic Dinoflagellate Blooms" (V. R. Locicero, ed.), p. 267. Mass Sci. Tech. Found, Wakefield, Massachusetts.

Scheuer, P. J. (1973). "Chemistry of Marine Natural Products." 201 pp. Academic Press, New York and London.

Scheuer, P. J. (1975). Recent developments in the chemistry of marine toxins. *Lloydia* **38**, 1.

Scheuer, P. J. (ed.) (1978). "Marine Natural Products. Chemical and Biological Perspectives", Vol. II. Academic Press, London and New York.

Scheuer, P. J. (ed.) (1980). "Marine Natural Products. Chemical and Biological Perspectives", Vol. III, 229 pp. Academic Press, London and New York.

Scheuer, P. J. (ed.) (1981). "Marine Natural Products. Chemical and Biological Perspectives", Vol. IV, 199 pp. Academic Press, London and New York.

Schwimmer, D. and Schwimmer, M. (1964). Medical aspects of phycology. *In* "Algae and Man" (D. F. Jackson, ed.), pp. 279–358. Plenum Press, New York.

Seville, R. H. (1957). Dogger bank itch. Report of a case. *Br. J. Derm* **69**, 92.

Shapiro, B. I. (1968a). Purification of a toxin from the tentacles of the anemone *Condylactis gigantea*. *Toxicon* **5**, 253.

Shapiro, B. I. (1968b). A site of action of toxin from the anemone *Condylactis gigantea*. *Comp. Biochem. Physiol.* **27**, 519.

Shapiro, B. I. and Lilleheil, G. (1969). The action of anemone toxin on crustacean neurons. *Comp. Biochem. Physiol.* **28**, 1225.

Sheumack, D. D., Howden, M. E. H., Spense, I. and Quinn, R. J. (1978). Maculotoxin: a neurotoxin from the venom glands of the octopus *Hapalochlaena maculosa* identified as tetrodotoxin. *Science* **199**, 188.

Shibota, M. and Hashimoto, Y. (1970). Purification of the ivory shell toxin. *Bull. Jpn. Soc. scient. Fish.* **36**, 115–119.

Shilo, M. and Rosenberger, R. F. (1960). Studies on the toxic principles formed by the chrysomonad *Prymnesium parvum* Carter. *Ann. N.Y. Acad. Sci.* **90**, 866.

Shimada, S. (1969). Antifungal steroid glycoside from sea cucumber. *Science* **163**, 1462.

Shimizu, Y. (1978). Dinoflagellate toxins. *In* "Marine Natural Products" (P. J. Scheuer, ed.), Vol. I, p. 1. Academic Press, New York and London.

Shimizu, Y., Alam, M., Oshima, Y. and Fallon, W. E. (1975). Presence of four toxins in red tide infested clams and cultured *Gonyaulax tamarensis* cells. *Biochem. Biophys. res. Commun.* **66**, 731.

Shin, L. M., Michaels, W. D. and Mayer, M. M. (1979). Membrane damage by a toxin from the sea anemone *Stoichactis helianthus*—II. Effect of membrane lipid composition in a liposome system. *Biochim. Biophys. Acta* **555**, 79.

Simon, S. E., Cairncross, K. D., Stachell, D. G., Gay, W. S. and Edwards, S. (1964). The toxicity of *Octopus maculosus* Hoyle venom. *Arch. intern. Pharmacodyn.* **148**, 318.

Sims, J. K. and VanRilland, R. D. (1981). Escharotic stomatitis caused by the "stinging seaweed" *Microcoleus lyngbyaceus*. *Hawaii med. J.* **40**, 243.

Sket, D., Drasler, K., Ferlan, I. and Lebez, D. (1974). Equinatoxin, a lethal protein from *Actinia equina*—II. Pathophysiological action. *Toxicon* **12**, 63.

Smith, F. G. W. (1954). Red tide studies. *Mar. Lab. Univ. Miami, Coral Gables* (*Prelim. Rep. to Fla. St. Bd. Conservation*), 117 pp.

Snow, C. D. (1970). Two accounts of the southern octopus, *Octopus dofleini*, biting scuba divers. *Oregon Fish. Comm. Res. Rep.* **2**, 103.

Solomon, A. E. and Stoughton, R. B. (1978). Dermatitis from purified sea algae toxin (debromoaplysiatoxin). *Arch. Dermatol.* **114**, 1333.

Sommer, H. (1932). The occurrence of the paralytic shell-fish poison in the common sand-crab. *Science* **76**, 574.

Sommer, H. and Meyer, K. F. (1937). Paralytic shellfish poisoning. *Arch. Path.* **24**, 560.

Sommer, H., Monnier, R. P., Riegel, B. *et al.* (1948). Paralytic shellfish poison. I. Occurrence and concentration by ion exchange. *J. Am. chem. Soc.* **70**, 1015.

Songdahl, J. H. and Shapiro, B. I. (1973). Purification and composition of a toxin from the posterior salivary gland of *Octopus dofleini*. *Toxicon* **12**, 109.

Southcott, R. V. (1979). Marine toxins. *In* "Handbook of Clinical Neurology" (M. H. Cohen and H. L. Klawans, eds), Vol. 37, pp. 27–106. North-Holland Publ. Co., Amsterdam.

Southcott, R. V. and Kingston, C. W. (1959). Lethal jellyfish stings: a study in sea-wasps. *Med. J. Aust.* (1), 443.

Spense, I. (1978). Long-lasting, reversible spastic paresis produced by a halogenated monoterpene isolated from the red alga *Plocanium carilagineum* (L.). *In* "Abst. 7th Internat. Congr. Pharmacol", Abst. 84.

Spense, I., Jamieson, D. D. and Taylor, K. M. (1979). Anticonvulsant activity of farnesylaclone expoxide—a novel marine product. *Experientia* **35**, 238–239.

Spiegelstein, M. Y., Paster, Z. and Abbott, B. C. (1973). Purification and biological activity of *Gymnodinium breve* toxins. *Toxicon* **11**, 85–93.

Starr, T. J. (1958). Notes on a toxin from *Gymnodinium brevis*. *Texas Rept. Biol. Med.* **20**, 271.

Steidinger, K. A. (1975). Basic factors influencing red tides. *In* "Toxic Dino-flagellate Blooms" (V. R. Locicero, ed.), pp. 152–162. Mass. Sci. Tech. Found., Wakefield, Massachusetts.

Stempien, M. F., Ruggieri, G. D., Nigrelli, R. F. and Cecil, J. T. (1970). Physiologically active substances from extracts of marine sponges. *In* "Food–Drugs from the Sea Conference, 1969" (H. W. Youngken, ed.), p. 295. Mar. Tech. Soc., Washington, D.C.

Stillway, L. W. and Lane, C. E. (1971). Phospholipase in the nematocyst toxin of *Physalia physalis*. *Toxicon* **9**, 193.

Sutherland, S. K. (1979). Response to *Chironex* antivenom. *Med. J. Aust.* **2**, 653.

Sutherland, S. K. and Lane. (1969). Toxins and mode of envenomation of the common ringed or blue-banded octopus. *Med. J. Aust.* **1**, 893–897.

Sutherland, S. K., Broad, A. J. and Lane, W. R. (1970). Octopus neurotoxins: low molecular weight non-immunogenic toxins present in the saliva of the blue-ringed octopus. *Toxicon* **8**, 249.

Tachibana, K. (1980). "Structural studies on marine toxins." Dissertation, University of Hawaii.

Tamkun, M. M. and Hessinger, D. A. (1981). Isolation and partial characterization of a hemolytic and toxic protein from the nematocyst venom of the Portuguese man-of-war, *Physalia physalis*. *Biochim. Biophys. Acta* **667**, 87–98; see also (1979) *Fed. Proc.* **38**, 824 (Abst.).

Tanaka, M., Haniu, M., Yasunobu, K. T. and Norton, T. R. (1977). Amino acid sequence of the *Anthopleura xanthogrammica* heart stimulant, Anthopleurin A. *Biochemistry* **16**, 204.

Tangen, K. (1979). Cited by Smayda, T. *In* "Toxic Dinoflagellate Blooms" (D. L. Taylor and H. H. Seliger, eds), p. 456. Elsevier/North-Holland, New York.

Taylor, D. L. and Seliger, H. H. (eds) (1979). "Toxic Dinoflagellate Blooms", 505 pp. Elsevier/North-Holland, New York.

Taylor, K. M. and Spense, I. (1979). Marine natural products affecting neurotransmission. *In* "Neurotoxins. Fundamental and Clinical Advances" (I. W. Chubb and L. B. Geffen, eds), pp. 85–93. Adelaide Univ. Union Press, Adelaide, S. Australia.

Teh, Y. F. and Gardiner, J. E. (1974). Partial purification of *Lophozozymus pictor* toxin. *Toxicon* **12**, 603–610.

Thompson, T. E. (1965). Epidermal acid-secretion in some marine polyclad turbellaria. *Nature* **206**, 954.

Thorpe, J. P. and Ryland, J. S. (1979). Cryptic speciation detected by biochemical genetics in three ecologically important intertidal bryozoans. *Eust. coast. mar. Sci.* **8**, 395–398.

Thron, C. D., Patterson, R. N. and Friess, S. L. (1963). Further biological properties of the sea cucumber toxin holothurin A. *Toxicol. appl. Pharmacol.* **5**, 1.

Thron, C. D., Durant, R. C. and Friess, S. L. (1964). Neuromuscular and cytotoxic effects of holothurin A and related saponins at low concentration levels. III. *Toxicol. appl. Pharmacol.* **6**, 182.

Togashi, M. (1943). [Chemical study of the poisoning by *Venerupis semidecussata*.] *Jpn. Iji Shimpo*, May 1943.

Tokunai (1790). Cited by Bisset, N. G. (1976). Hunting poisons of the North Pacific region. *Lloydia* **39**, 87.

Tomlin, E. W. F. (1966). The Ainu: their history and culture. *J. Roy. Cent. Asia Soc.* **53**, 297.

Trethewie, E. R. (1965). Pharmacological effects of the venom of the common octopus *Hapalochlaena maculosa*. *Toxicon* **3**, 55.

Trevan, J. W. (1927). The error of determination of toxicity. *Proc. Roy. Soc.* (London) **B101**, 483.

Trieff, N. M., Venkatas, N. and Ray, S. M. (1972). Purification of *Gymnodinium breve* toxin—dry column chromatographic techniques. *Tex. J. Sci.* **23**, 596.

Tsutsumi, J. and Hashimoto, Y. (1964). Isolation of pyopheophorbide *a* as a photodynamic pigment from the liver of abalone *Haliotis discus hannai*. *Agric. Biol. Chem.* **28**, 467–470.

Turk, J. L., Parker, D. and Rudner, E. J. (1966). Preliminary results on the purification of the chemical sensitizing agent in *Alcyonidium galathinosum*. *Proc. Roy. Soc. Med.* **59**, 1122–1124.

Turlapaty, P., Shibata, S., Norton T. R. and Kashiwagi, M. (1973). A possible mechanism of action of a central stimulant substance isolated from sea anemone *Stoichactis kenti*. *Eur. J. Pharmac.* **24**, 310.

Tweedie, M. W. F. (1941). "Poisonous Animals of Malaya." 90 pp. Singapore Publ. House, Singapore.

Uemura, D., Ueda, K., Hirata, Y., Katayama, C. and Tanaka, J. (1980). Structural studies on palytoxin, a potent coelenterate toxin. *Tetrahedron Lett.* **21**, 4867–4871.

Ulitzur, S. and Shilo, M. (1970). Effects of *Prymnesium parvum* toxins, cetyl trimethylammonium bromide and sodium dodecyl sulphate on bacteria. *J. gen. Microbiol.* **62**, 363–370.

Von Uexkull, J. (1899). Die Physiologie der Pedicellarien. *Z. Biol. Ser.* 2 **37**, 344.

Wall, D. (1975). Taxonomy and cysts of red-tide dinoflagellates. *In* "Toxic Dinoflagellate Blooms" (V. R. Locicero, ed.), pp. 249–255. Mass. Sci. Tech. Found., Wakefield, Massachusetts.

Wang, C. M., Narahashi, T. and Mendi, T. J. (1973). Depolarizing action of *Haliclona* toxin on end-plate and muscle membranes. *Toxicon* **11**, 499.

Wangersky, P. J. and Guillard, R. R. L. (1960). Low molecular weight organic base from the dinoflagellate *Amphidinium carteri*. *Nature* **185**, 689.

Waraskiewicz, S. M. and Erickson, K. L. (1974). Halogenated sesquiterpenoids from the Hawaiian marine alga *Laurencia nidifica:* nidificene and nidifidiene. *Tetrahedron Lett.* No. 23, 2003–2006.

Warnick, J. E., Weinreich, D. and Burnett, J. W. (1981). Sea nettle (*Chrysaora quinquecirrha*) toxin on electrogenic and chemosensitive properties of nerve and muscle. *Toxicon* **19**, 361–371.

Watson, M. (1973). Midgut gland toxins of Hawaiian sea hares—I. Isolation and preliminary toxicology observations. *Toxicon* **11**, 259–267.

Watson, M. and Rayner, M. D. (1973). Midgut gland toxins of Hawaiian sea hares— II. A preliminary pharmacological study. *Toxicon* **11**, 269–276.

Weill, R. (1934). "Contributions a l'étude des Cnidaires et de leur Nematocystes." Trav. Stat. Zool. Wimiereux, Paris, Tome 10, 11, 347 pp.

Welsh, J. H. and Moorhead, M. (1960). The quantitative distribution of 5-hydroxytryptamine in the invertebrates, especially in their nervous system. *J. Neurochem.* **6**, 146.

Werner, B. (1965). Die Nesselkapseln der Cnidaria mit besonderen Berücksichtigung der Hydroida. I. Klassifikation und Bedeutung für die Systematik und Evolution. *Helgol. Wiss. Meeresunters* **12**, 1–39.

White, A. W. (1977). Dinoflagellate toxins as probable cause of an Atlantic herring (*Clupea harengus harengus*) kill, and pteropods as an apparent vector. *J. Fish. Res. Bd. Can.* **34**, 2421–2424.

Whittaker, V. P. (1960). Pharmacologically active choline esters in marine gastropods. *Ann. N.Y. Acad. Sci.* **90**, 695.

Whittaker, V. P. and Wijesuneera, S. (1952). The separation of esters of choline by filter-paper chromatography. *Biochem. J.* **51**, 348.

Whysner, J. A. and Saunders, P. R. (1966). Purification of the lethal fraction of the venom of the marine snail *Conus californicus*. *Toxicon* **4**, 177.

Whyte, J. M. and Endean, R. (1962). Pharmacological investigation of the venoms of the marine snails *Conus textile* and *Conus geographus*. *Toxicon* **1**, 25.

Wiberg, G. S. and Stephenson, N. R. (1960). Toxicologic studies on paralytic shellfish poison. *Toxicol. appl. Pharmacol.* **2**, 607–615.

Wiles, J. S., Vick, J. A. and Christensen, M. K. (1974). Toxicological evaluation of palytoxin in several animal species. *Toxicon* **12**, 427.

Winkler, L. R. (1961). Preliminary tests of the toxin extracted from California sea hares of the genus *Aplysia*. *Pac. Sci.* **15**, 211.

Winkler, L. R., Tilton, B. E. and Hardinge, M. G. (1962). A cholinergic agent extracted from sea hares. *Arch. intern. Pharmacodyn.* **137**, 76–83.

Wittle, L. W. and Wheeler, C. A. (1974). Toxic and immunological properties of stinging coral toxin. *Toxicon* **12**, 487–493.

Wittle, L. W., Middlebrook, R. E. and Lane, C. E. (1971). Isolation and partial purification of a toxin from *Millepora alcicornis*. *Toxicon* **9**, 327–331.

Wong, J. L., Oesterlin, R. and Rapoport, H. (1971). The structure of saxitoxin. *J. Am. chem. Soc.* **93**, 7344.

Woodward, G. (1955). "Conference on Shellfish Toxicology", p. 26. Public Health Service, Washington, D.C.

Yamamura, S. and Hirata, Y. (1963). Structures of aplysin and aplysinol, naturally occurring bromo-compounds. *Tetrahedron* **19**, 1485–1496.

Yamanouchi, T. (1929). Notes on the behavior of the holothurian *Caudina chilensis* (J. Muller). *Sci. Rep. Res. Ints. Tohoku Univ.*, *(Biol.)* **4**, 73–115.

Yamanouchi, T. (1942). Study of poisons contained in holothurians. *Teikoku Gakushiin Hokoku* **17**, 73.

Yamanouchi, T. (1955). On the poisonous substance contained in Holothurians. *Publ. Seto Mar. Biol. Lab.* **4**, 184.

Yanagita, T. M. (1959a). Physiological mechanism of nematocyst responses in sea anemone II. Effects of electrolyte ions upon the isolated cnide. *J. Fac. Sci. Univ. Tokyo, Sect. IV* **8**, 381–400.

Yanagita, T. M. (1959b). Physiological mechanism of nematocyst responses in sea anemone VII. Extrusion of resting cnidae—its nature and its possible bearing on the normal nettling response. *J. Exp. Biol.* **36**, 478–494.

Yanagita, T. M. (1973). The "cnidoblast" as an excitable system. *In* "Recent Trends in Research in Coelenterate Biology." (T. Tokioka and S. Nishimura, eds), Vol. 20 p. 675. Publ. Seto Mar. Biol. Lab., Shirahama, Wakayama-Kem, Japan.

Yariv, J. and Hestrin, S. (1961). Toxicity of the extracellular phase of *Prymnesium parvum* cultures. *J. gen. Microbiol.* **24**, 165.

Yasumoto, T. and Hashimoto, Y. (1965). Properties and sugar components of asterosaponin A isolated from starfish. *Agric. biol. Chem.* **31**, 368.

Yasumoto, T., Watanabe, T. and Hashimoto, Y. (1964). [Physiological activities of starfish saponin.] *Bull. Jpn. Soc. scient. Fish.* **30**, 357.

Yasumoto, T., Tanaka, M. and Hashimoto, Y. (1966). Distribution of saponin in echinoderms. *Bull. Jpn. Soc. scient. Fish.* **32**, 673.

Yasumoto, T., Nakumura, K. and Hashimoto, Y. (1967). A new saponin, holothurin B, isolated from sea-cucumber, *Holothuria vagabunda* and *Holothuria lubrica*. *Agric. biol Chem.* **31**, 7.

Yentsch, C. M. (1981). Flow cytometric analysis of cellular saxitoxin in the dinoflagellate *Gonyaulax tamarensis* var. *excavata*. *Toxicon* **19**, 611–621.

Zervos, S. G. (1934). La maladie des pêcheurs d'éponges. *Paris Med.* **93**, 89–97.

# Taxonomic Index

# Subject Index

(F = Figure; T = Table)

227

# Cumulative Index of Titles

231

# Cumulative Index of Authors